Ben Stacy Jerrik (Ed.)

Maryland Route 23

Ben Stacy Jerrik (Ed.)

Maryland Route 23

U.S. Route 1 in Maryland, Pennsylvania Route 24, Maryland Route 165, Maryland Route 146

Part Press

Contents

Articles

References

Maryland_Route_23

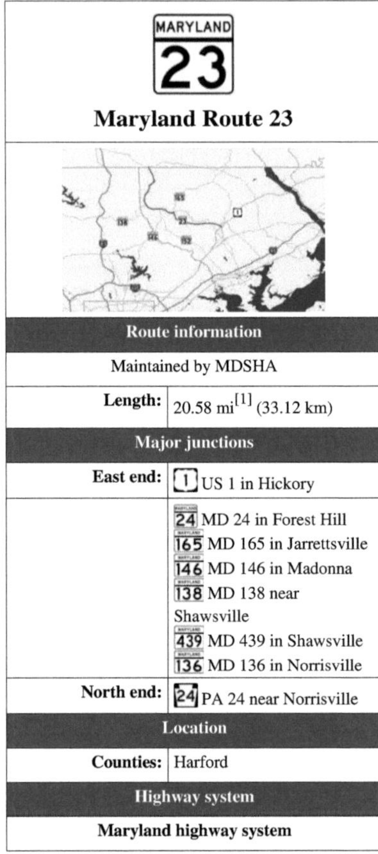

Maryland Route 23 (**MD 23**) is a state highway in the U.S. state of Maryland. The state highway runs 20.58 miles (33.12 km) from U.S. Route 1 (US 1) in Hickory north and west to the Pennsylvania state line near Norrisville, where the highway continues as Pennsylvania Route 24 (PA 24). MD 23 is an L-shaped highway in northwestern Harford County that consists of two major sections. Between US 1 and MD 165 in Jarrettsville, MD 23 is marked east–west along a two-lane controlled access road named **East–West Highway**. From MD 165 to the state line, the state highway is marked north–south along Norrisville Road, a rural two-lane highway that passes through the villages of Madonna and Shawsville. The two sections of MD 23 are connected by a short concurrency with MD 165.

MD 23 from Hickory to Jarrettsville was one of the original state highways marked for improvement by the Maryland State Roads Commission in 1909 and one of the original state-numbered highways in 1927. The state road was constructed from Hickory to Jarrettsville and from Norrisville to the state line in the early to mid-1910s. The gaps between Jarrettsville and Shawsville and from Shawsville to Norrisville were filled in the early 1920s. East–West Highway was constructed in the early 1960s to replace the parallel Jarrettsville Road. MD 23's eastern terminus was moved to US 1's new bypass of Hickory in 2000.

Route description

MD 23 begins at an intersection with US 1 (Hickory Bypass) in Hickory. The state highway heads west as East–West Highway, a two-lane controlled access highway that parallels the old alignment of MD 23, Jarrettsville Road. MD 23 passes through a commercial area where the highway intersects US 1 Business (Conowingo Road) and Water Tower Way before the highway passes through a narrow forested corridor. The state highway intersects Commerce Road at a roundabout; Commerce Road heads north into an industrial park surrounding Forest Hill Airport. After crossing over Bynum Run and the Ma and Pa Trail, a rail trail along the abandoned right-of-way of the Maryland and Pennsylvania Railroad, MD 23 intersects MD 24 (Rocks Road/Rock Springs Avenue) just south of Forest Hill. MD 23 continues west through a mix of farmland and forest, passing south of the hamlet of Fairview and crossing over Phillips Mill Road and Morse Road before East–West Highway reaches its western terminus at MD 165 (Baldwin Mill Road) south of Jarrettsville.[1] [2]

MD 23 turns north and joins MD 165 in a concurrency to the center of Jarrettsville. MD 165 continues north as Federal Hill Road and MD 23 joins its original alignment, which heads east as Jarrettsville Road. MD 23 heads west through farmland as Norrisville Road, where the highway intersects MD 146 (Jarrettsville Pike) and Madonna Road in the village of Madonna. The state highway gradually curves to the north through Shawsville, where MD 23 intersects two highways that head west into Baltimore County, MD 138 (Troyer Road) and MD 439 (Old York Road). The state highway makes a sharp, curvaceous descent to a crossing of Deer Creek, next to which is the historic Ivory Mills complex. MD 23 ascends out of the forested creek valley into Norrisville and meets the north end of MD 136 (Harkins Road). The state highway continues north to just south of the Pennsylvania state line. MD 23 turns west and closely parallels the state line for about 2000 feet (610 m) before the highway veers northwest to cross the border. The highway continues north as PA 24 (Barrens Road) toward Stewartstown.[1] [2]

History

The first section of MD 23 to be paved was the part of the Old York Turnpike from MD 138 to MD 439, which was improved as a 12-foot (3.7 m) wide macadam state-aid road by 1910.[3] [4] The portion of MD 23 from Hickory to Jarrettsville was designated one of the original state roads by the Maryland State Roads Commission in 1909.[3] The state road was constructed with a 14-foot (4.3 m) wide macadam surface from Hickory to Grafton Shop Road in 1910, from Morse Road to Jarrettsville in 1914, and between Grafton Shop Road and Morse Road in 1915.[5] Another section of 12-foot (3.7 m) wide macadam road was constructed under state aid from south of Norrisville to the Pennsylvania state line by 1915.[5] The highway from Jarrettsville to MD 138 was under construction as a concrete road by 1919 and completed in 1921.[6] [7] The final portion of MD 23, also paved in concrete, was completed from MD 439 to south of Norrisville in 1923.[8] MD 23 was one of the state numbered highways assigned in 1927.[9] The state highway was widened to a width of 20 feet (6.1 m) from Hickory to Jarrettsville by 1930.[10]

Jarrettsville Road was replaced by East–West Highway, which was under construction by 1961 and completed in 1963.[11] [12] The MD 23 designation was moved to the new highway and Jarrettsville Road was immediately transferred to county maintenance.[12] At its eastern end, East–West Highway was constructed to tie into the Bel Air Bypass that was completed in 1965 and was proposed to continue to east of Hickory.[13] Close to its eastern end, MD 23 curved south to an intersection with Granary Road. MD 23 turned east onto Granary Road for its eastern terminus at US 1 (now US 1 Business). East–West Highway continued south as MD 23A; at Bynum Road, MD 23A became a one-lane ramp that joined the Bel Air Bypass.[14] [15] In 2000, the Hickory Bypass was completed; US 1 was moved to the new bypass and US 1 Business was extended along US 1's old route through Hickory. As part of the same project, MD 23 was extended east to an new terminus at the Hickory Bypass. East–West Highway's curve to the south was replaced with a perpendicular intersection with MD 23A, which was renamed Water Tower Way, extended north to new MD 23, and gained a ramp from northbound US 1 at its southern end. The portion of Granary Road that formed MD 23's terminus was designated MD 23B.[16] Both MD 23A and MD 23B were transferred to Harford County maintenance in 2002.[17] MD 23's roundabout at Commerce Road in Forest Hill was constructed in

2008.[18]

Junction list

MD 23 is signed east–west east of MD 165 and north–south west of MD 165. The mileage below starts from the eastern terminus. The entire route is in Harford County.

Location	Mile [1]	Destinations	Notes
Hickory	0.00	US 1 (Hickory Bypass) – Baltimore, Rising Sun	Eastern terminus
	0.22	US 1 Bus. (Conowingo Road) – Bel Air	
Forest Hill	1.12	Commerce Road north	Roundabout
	2.00	MD 24 (Rocks Road/Rock Spring Avenue) – Bel Air, Pylesville	
Jarrettsville	6.81	MD 165 south (Baldwin Mill Road) – Baldwin	South end of concurrency with MD 165
	8.09	MD 165 north (Federal Hill Road) / Jarrettsville Road east – Pylesville, Forest Hill	North end of concurrency with MD 165; Jarrettsville Road is old alignment of MD 23
Madonna	10.44	MD 146 south (Jarrettsville Pike) / Madonna Road north – Jacksonville	
Shawsville	12.66	MD 138 west (Troyer Road) / Troyer Road east – Monkton	
	13.88	MD 439 west (Old York Road) – Maryland Line	
Norrisville	18.58	MD 136 east (Harkins Road) – Whiteford	
	20.58	PA 24 (Barrens Road) – Stewartstown	Northern terminus; Pennsylvania state line
1.000 mi = 1.609 km; 1.000 km = 0.621 mi			

Auxiliary routes

MD 23 previously had two auxiliary routes.

- MD 23A was the designation for the portion of East–West Highway between Granary Road and the ramp to southbound US 1 (Bel Air Bypass). The 0.38-mile (0.61 km) state highway was constructed in the mid 1960s as a connection between MD 23 and southbound US 1.[15] In 2000, MD 23A was extended north to the new MD 23 and a ramp from northbound US 1 to MD 23A was added. The highway had a new length of 0.53 miles (0.85 km) and was renamed Water Tower Way.[16] [19] MD 23A was removed from the state highway system in 2002.[17]
- MD 23B was the designation for the 0.16-mile (0.26 km) section of Granary Road between US 1 Business and MD 23A (Water Tower Way) that was part of MD 23 from the construction of East–West Highway in the early 1960s until 2000.[16] [20] MD 23B was assigned in 2000 and removed from the state system in 2002.[16] [17]

References

[1] "Highway Location Reference: Harford County" (http://www.marylandroads.com/Location/2009_HARFORD.pdf) (PDF). Maryland State Highway Administration. 2009. . Retrieved 2011-02-23.

[2] Google, Inc. *Google Maps − Maryland Route 23* (http://maps.google.com/maps?f=d&source=s_d&saddr=MD-23+W&daddr=MD-23+S/Norrisville+Rd+to:MD-23+N/Norrisville+Rd&hl=en&geocode=FcDJWwIdQPty-w;Fbx2XAIdLB1w-w;FScZXgIdMuxv-w&mra=ls&sll=39.608135,-76.527414&sspn=0.014944,0.038581&ie=UTF8&t=h&z=11) (Map). Cartography by Google, Inc. . Retrieved 2011-02-23.

[3] Maryland Geological Survey. *Map of Maryland* (Map) (1910 ed.).

[4] United States Geological Survey. *Parkton, MD quadrangle* (http://historical.mytopo.com/quad.cfm?quadname=Parkton&state=MD&series=15) (Map). 1:48,000. 15 Minute Series (Topographic) (1938 ed.). . Retrieved 2011-02-24.

[5] *Report of the State Roads Commission of Maryland* (http://www.archive.org/details/annualreportsofs1912mary). **1912-1915**. Baltimore: Maryland State Roads Commission. May 1915. pp. 112, 124. . Retrieved 2011-02-24.

[6] *Report of the State Roads Commission of Maryland* (http://www.archive.org/details/annualreportsofs1916mary). **1916-1919**. Baltimore: Maryland State Roads Commission. January 1920. p. 39. . Retrieved 2011-02-24.

[7] Maryland Geological Survey. *Map of Maryland: Showing State Road System and State Aid Roads* (Map) (1921 ed.).

[8] Maryland Geological Survey. *Map of Maryland: Showing State Road System and State Aid Roads* (Map) (1923 ed.).

[9] Maryland Geological Survey. *Map of Maryland: Showing State Road System and State Aid Roads* (Map) (1927 ed.).

[10] *Report of the State Roads Commission of Maryland* (http://www.archive.org/details/reportofstateroa1927mary). **1927-1930**. Baltimore: Maryland State Roads Commission. 1930-10-01. p. 82. . Retrieved 2011-02-24.

[11] "NBI Structure Number: 100000120064010" (http://nationalbridges.com/nbi_record.php?StateCode=24&struct=100000120064010). *National Bridge Inventory*. . Retrieved 2011-02-24.

[12] Maryland State Roads Commission. *Maryland: Official Highway Map* (Map) (1963 ed.).

[13] Maryland State Roads Commission. *Maryland: Official Highway Map* (Map) (1965 ed.).

[14] United States Geological Survey (1986-07-01). *26 km NE of Baltimore, Maryland, United States* (http://msrmaps.com/image.aspx?T=2&S=12&Z=18&X=479&Y=5475&W=1&qs=) (Map). Topo Map. . Retrieved 2011-02-24.

[15] "Highway Location Reference: Harford County" (http://www.marylandroads.com/Location/1999_HARFORD.pdf) (PDF). Maryland State Highway Administration. 1999. . Retrieved 2011-02-24.

[16] "Highway Location Reference: Harford County" (http://www.marylandroads.com/Location/2000_HARFORD.pdf) (PDF). Maryland State Highway Administration. 2000. . Retrieved 2011-02-24.

[17] "Highway Location Reference: Harford County" (http://www.marylandroads.com/Location/2002_HARFORD.pdf) (PDF). Maryland State Highway Administration. 2002. . Retrieved 2011-02-24.

[18] "Highway Location Reference: Harford County" (http://www.marylandroads.com/Location/2008_HARFORD.pdf) (PDF). Maryland State Highway Administration. 2008. . Retrieved 2011-02-24.

[19] Google, Inc. *Google Maps − Maryland Route 23A* (http://maps.google.com/maps?f=d&source=s_d&saddr=East-West+Hwy&daddr=East-West+Hwy&hl=en&geocode=FUjNWwIdPOJy-w;FRWxWwId295y-w&mra=ls&sll=39.56471,-76.357555&sspn=0.007477,0.01929&ie=UTF8&ll=39.567174,-76.357491&spn=0.007477,0.01929&t=h&z=16) (Map). Cartography by Google, Inc. . Retrieved 2011-02-24.

[20] Google, Inc. *Google Maps − Maryland Route 23B* (http://maps.google.com/maps?f=d&source=s_d&saddr=East-West+Hwy&daddr=Granary+Rd&hl=en&geocode=FS7EWwIdHuJy-w;FRrEWwIdkO1y-w&mra=ls&sll=39.569126,-76.355839&sspn=0.003738,0.009645&ie=UTF8&ll=39.568175,-76.355635&spn=0.003738,0.009645&t=h&z=17) (Map). Cartography by Google, Inc. . Retrieved 2011-02-24.

External links

- MD 23 @ MDRoads.com (http://www.mdroads.com/routes/022-039.html#md023)

U.S._Route_1_in_Maryland

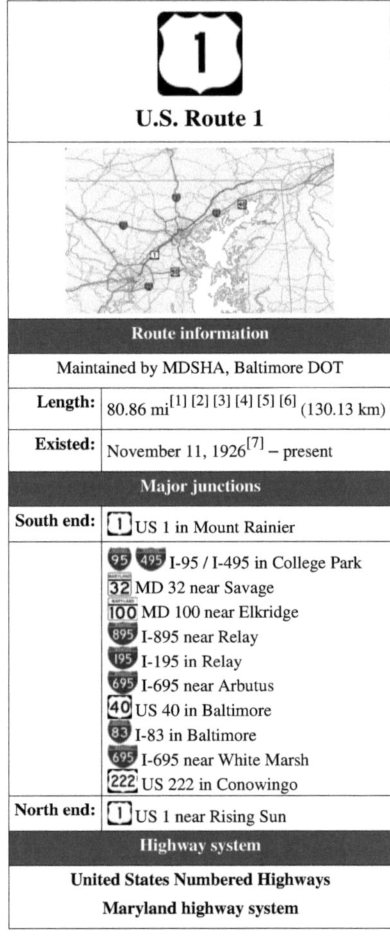

U.S. Route 1

Route information	
Maintained by MDSHA, Baltimore DOT	
Length:	80.86 mi[1] [2] [3] [4] [5] [6] (130.13 km)
Existed:	November 11, 1926[7] – present
Major junctions	
South end:	US 1 in Mount Rainier
	I-95 / I-495 in College Park MD 32 near Savage MD 100 near Elkridge I-895 near Relay I-195 in Relay I-695 near Arbutus US 40 in Baltimore I-83 in Baltimore I-695 near White Marsh US 222 in Conowingo
North end:	US 1 near Rising Sun
Highway system	
United States Numbered Highways **Maryland highway system**	

U.S. Route 1 (**US 1**) is the easternmost and longest of the major north–south routes of the United States Numbered Highway System, running from Key West, Florida to Fort Kent, Maine. In the U.S. state of Maryland, it is an 80.86-mile (130.13 km) segment of the route that runs through central Maryland between Mount Rainier and Rising Sun.

US 1 is paralleled by several major highways as it passes through Maryland, including Interstate 95, the Baltimore–Washington Parkway, U.S. Route 29 and U.S. Route 301. Thus, US 1 has lost its significance as a long distance route through the state. It is often congested, however, because it remains a major route in the individual towns it traverses.

Route description

Prince George's County

US 1 leaves the District of Columbia and enters Maryland at the town of Mount Rainier. The highway heads northeast from Eastern Avenue as Rhode Island Avenue, a four-lane divided street with parking through a downtown-like commercial area. US 1 meets 34th Avenue and Perry Street at a roundabout, then continues northeast through a densely-populated residential area. The highway leaves Mount Rainier and enters Brentwood, where the highway meets MD 208 (38th Street). US 1 passes through North Brentwood as a four-lane divided highway without parking through a mix of residences and commercial establishments. The median widens as the highway crosses Northwest Branch and enters the town of Hyattsville. The highway begins to closely

US 1 approaching Madison Street in Hyattsville

parallel CSX's Capital Subdivision and MARC's Camden Line as it reduces to a four-lane undivided highway, passing Prince George's District Court. US 1 curves to the north and the highway's name changes to Baltimore Avenue at Farragut Street, shortly before intersecting US 1 Alternate.

US 1 continues north through downtown Hyattsville, gaining a center turn lane before entering the town of Riverdale Park, where the highway intersects MD 410 (East–West Highway). The highway enters a densely-populated residential area, passing between Riverdale Park to the east and the town of University Park to the west. Shortly after the highway enters the town of College Park on the east, US 1 intersects Queens Chapel Road at a five-way intersection. Only buses may enter Queens Chapel Road from US 1. The highway fully enters College Park and enters the commercial area that makes up the downtown of the college town, expanding to a four-lane divided highway. After the intersection with College Avenue and Regents Drive, US 1 passes through the campus of the University of Maryland, College Park, including the historic Rossborough Inn and Fraternity Row. The highway leaves the campus after intersecting Campus Drive and Paint Branch Parkway and crossing Paint Branch.

US 1 continues through the northern part of College Park as a five-lane road with center turn lane, passing through a suburban commercial area. The highway intersects Greenbelt Road, which is unsigned MD 430, before meeting MD 193 (University Boulevard) at a partial interchange. All movements not provided in the interchange require using Greenbelt Road to connect between US 1 and MD 193. After intersecting Cherry Hill Road, US 1 becomes a divided highway and meets the Interstate 495 (Capital Beltway) (I-95 and I-495) at a partial cloverleaf interchange. North of the interchange, the highway expands to a six-lane divided highway, leaving the town of College Park and passing by an Ikea store and through a swath of the Beltsville Agricultural Research Center, including the National Agricultural Library.

At Sunnyside Avenue, the highway reduces to a five-lane road with center turn lane and passes through a suburban commercial area in the unincorporated town of Beltsville. The highway meets Rhode Island Avenue at an oblique intersection and MD 212 (Powder Mill Road) at a more orthogonal intersection before curving north and paralleling CSX's Capital Subdivision. The highway passes between a commercial strip on the southbound side of the highway and an industrial area on the east side of the railroad tracks. US 1 leaves Beltsville after crossing Indian Creek. The highway temporarily expands to a four-lane divided highway before returning to a five-lane road, passing between an industrial area to the east and office parks to the west. After Miurkirk Meadows Drive, which leads to Miurkirk Road, US 1 reduces to a four-lane undivided highway, passes under the latter highway, and enters a forested area.

US 1 gradually veers away from the railroad tracks as it approaches Laurel, passing Maryland National Memorial Park before entering a suburban commercial area ahead of Contee Road, where the center turn lane returns. At Cypress Street, the southbound direction gains a third lane through the intersection with Cherry Lane, where the northbound direction gains a third lane. After passing Laurel Mall and Laurel Shopping Center, US 1 splits into a one-way pair. The three to four northbound lanes veer northeast as Second Street, while the three to four southbound lanes take the name Washington Boulevard. The one-way pair intersects Bowie Road, the old alignment of MD 197, before intersecting MD 198, which takes the form of a one-way pair, eastbound Gorman Avenue and westbound Talbott Avenue. US 1 continues through the city of Laurel, intersecting Main Street just west of the Laurel MARC Station before leaving Laurel by crossing the Patuxent River into Howard County.

Howard County

Immediately after crossing the Patuxent River, both directions of US 1 pass entrances to Laurel Park Racecourse. The highway continues through North Laurel, with the one-way pair coming together shortly before the intersection with Whiskey Bottom Road. US 1 continues north as a five-lane road with center turn lane, crossing Hammond Branch. As the highway passes through a commercial area in Savage, the road becomes a divided highway. After Gorman Road, the roadways temporarily diverge to cross the Little Patuxent River, then come together again at the full cloverleaf interchange with MD 32 (Patuxent Freeway). US 1 becomes undivided and intersects Guilford Road and crosses over the Columbia Branch from the CSX Capital Subdivision to the east. The highway passes several industrial parks, crossing Dorsey Run twice before intersecting MD 175 (Waterloo Road) in Jessup.

North of Jessup, US 1 crosses Deep Run and passes more industrial parks. The highway intersects MD 103 (Meadowridge Road) and passes Meadow Ridge Memorial Park before a partial cloverleaf interchange with MD 100 near Dorsey. Continuing northeast past Bealmear Branch, Old Washington Road splits to the northeast to pass through the center of Elkridge shortly before US 1 intersects the old alignment of MD 103, Montgomery Road. The highway passes through a shallow S-curve before receiving the other end of Old Washington Road and passes under the Capital Subdivision. US 1's two directions become divided by a Jersey barrier as the highway passes through the commercial strip in Elkridge, passing ramps to and from I-895 (Harbor Tunnel Thruway) immediately before crossing the Patapsco River into Baltimore County.

Baltimore City and County

After passing under I-895, US 1 intersects South Street, which provides access to St. Denis and Relay, and passes through its interchange with I-195 (Metropolitan Boulevard). The undivided highway crosses over CSX's Baltimore Terminal Subdivision before reaching the partial interchange with US 1 Alternate. Washington Boulevard continues straight northeast as US 1 Alternate through Halethorpe, while US 1 exits onto Southwestern Boulevard to head north through Arbutus. The three-lane road with center turn lane and extra-wide shoulders parallels the Amtrak Northeast Corridor and MARC's Penn Line. After crossing Herbert Run, US 1 parallels the long, linear parking lot of the Halethorpe MARC station. The highway leaves the station after passing under Francis Avenue. US 1 expands to

View of North Avenue in Baltimore, looking east at the intersection of Barclay Street

a five-lane road with center turn lane as it crosses under I-95. US 1 passes under Sulphur Spring Road and I-695 (Baltimore Beltway) as the highway heads out of Arbutus. The only connection with I-695 is a single ramp, Exit 12A, from I-695 east to US 1 south. Inside the Beltway, US 1 is paralleled by Leeds Avenue, which intersects the federal highway just before entering the city of Baltimore.

US 1 continues north as a four-lane divided street through a densely-populated residential area. A short distance north of the city line, US 1 curves to the east onto Wilkens Avenue, which continues west as MD 372. The highway crosses the Amtrak Northeast Corridor and reduces to an undivided highway, intersecting US 1 Alternate (Caton Avenue) as it passes through a curve. US 1 crosses Gwynns Falls and expands to a divided highway before passing the Deck of Cards rowhouses between Brunswick Street and Millington Avenue. A short distance to the east, US 1 turns north onto a one-way pair, Monroe Street southbound and Fulton Avenue northbound, to pass through a densely-populated residential area on the west side of central Baltimore. US 1 meets the eastern terminus of MD 144 at another one-way pair, Pratt Street and Lombard Street. The highway passes over the US 40 freeway, which is accessed by the one-way pair adjacent to the freeway, Mulberry Street and Franklin Street. At Edmondson Avenue, Fulton Avenue becomes two-way; however, southbound US 1 remains on Monroe Street. Shortly after passing over the Amtrak Northeast Corridor, US 1 reaches North Avenue. The Fulton Avenue/Monroe Street one-way pair continues north as MD 140, while both directions of US 1 turn east onto North Avenue.

North Avenue heads east as a four-lane undivided highway through the junction with MD 129, which follows southbound Druid Hill Avenue and northbound McCulloh Street. US 1 continues east as a four-lane divided highway with parking lanes. The highway meets I-83 (Jones Falls Expressway) and Mount Royal Avenue next to the North Avenue station of the Baltimore Light Rail. US 1 expands to six lanes and crosses over the light rail tracks, Jones Falls, Falls Road, and CSX's Baltimore Terminal Subdivision. Due north of downtown Baltimore, North Avenue intersects Charles Street (MD 139 to the north), St. Paul and Calvert Streets (MD 2 to the south), and Greenmount Avenue (MD 45 to the north), after which North Avenue passes along the north side of Greenmount Cemetery. North Avenue reduces to a four-lane undivided street east of MD 45. Shortly after intersecting MD 147 (Harford Road) and Broadway, US 1 turns northeast onto Bel Air Road.

US 1 heads northeast as a four-lane undivided road, crossing the Baltimore Terminal Subdivision again and passing Clifton Park before reaching Erdman Avenue, which heads southeast as MD 151. The highway crosses Herring Run before intersecting Moravia Road, where Bel Air Road transitions from passing through densely-populated residential neighborhoods to being the center of a commercial strip. US 1 intersects Frankford Avenue and Hamilton Avenue before leaving the city of Baltimore at Fleetwood Avenue, which forms the eastbound component of a one-way pair with Northern Parkway at the latter street's eastern terminus. US 1 passes through the inner suburb of Overlea, crossing Stemmers Run before meeting I-695 (Baltimore Beltway) at a cloverleaf interchange.

US 1 continues northeast from the Beltway as a six-lane divided highway. The highway intersects Rossville Boulevard and Putty Hill Road then crosses MD 43 (White Marsh Boulevard). US 1 accesses MD 43 via connectors with a right-in/right-out interchange southbound and an intersection with Dunfield Road northbound. North of MD 43, US 1 reduces to a four-lane undivided highway as it passes through Perry Hall, intersecting Silver Spring Road, Joppa Road, and Ebenezer Road in the midst of a commercial strip. After the intersection with Honeygo Boulevard, US 1 descends into a steep valley to cross Gunpowder Falls. The highway continues north through a mix of farmland and residential subdivisions, heading through Kingsville before passing through Gunpowder Falls State Park and leaving Baltimore County by crossing Little Gunpowder Falls.

Harford County

US 1 heads northeast into Harford County, passing through a forested area and then the residential subdivisions of Pleasant Hills before reaching MD 152 (Mountain Road). The highway continues east through a commercial strip toward the hamlet of Benson, where US 1 converges with MD 147 (Harford Road) at an acute angle. US 1 Business heads east from the intersection toward Bel Air, while US 1 continues northeast on the Bel Air Bypass. The bypass begins as a four-lane divided highway, but drops to a two-lane road before crossing Winters Run. US 1 intersects Tollgate Road, gaining a lane northbound, before meeting MD 24 (Vietnam Veterans Memorial Highway) at a T-intersection with sweeping right-turn ramps. MD 24 joins US 1 in a concurrency north on a four-lane undivided highway that passes under Vale Road. MD 24 leaves US 1 at a partial cloverleaf interchange with Rock Spring Avenue, which heads south as MD 924 toward downtown Bel Air. The bypass continues northeast with two lanes, its name changed to the Hickory Bypass before expanding to a four-lane divided highway and intersecting US 1 Business (Conowingo Road) south of Hickory. US 1 intersects the eastern end of MD 23 (East–West Highway) and MD 543 (Ady Road) before reducing to a two-lane road and receiving the northern end of US 1 Business.

US 1 continues northeast as Conowingo Road through farmland. The old alignment of US 1, Forge Hill Road, splits to the east in Kalmia and rejoins the present highway 3 miles (4.8 km) later after separate crossings of Deer Creek. The highway passes through a commercial strip around the intersectin with MD 136 (Whiteford Road/Priestford Road) as US 1 passes to the south of Dublin. US 1 curves east at the intersection with MD 440 (Dublin Road). Smith Road, another old alignment of US 1, splits to the northeast as US 1 heads toward Darlington, where it meets MD 161 (Main Street) and MD 623 (Castleton Road). The highway heads straight over a series of hills, gaining climbing lanes in either direction. At Shuresville Road, US 1 curves to the northeast and crosses the Susquehanna River on top of Conowingo Dam.

Cecil County

On the east side of the river, US 1 passes over Norfolk Southern Railway's Columbia and Port Deposit Branch and immediately intersects MD 222 (Susquehanna River Road). The highway gains a climbing northbound as it ascends a hill to Conowingo, where the highway meets the southern terminus of US 222 (Rock Springs Road). US 1 turns east and meets a pair of old alignments: Connelly Road heading northwest and MD 591 (Colora Road) heading southeast toward a removed bridge over Octoraro Creek. After US 1 crosses the creek, the old alignment reconnects with the mainline as another segment of MD 591, Porters Bridge Road. West of Rising Sun, MD

US 1 northbound in Conowingo along concurrency with Maryland Route 222 Truck.

273 (Rising Sun Road) continues straight toward the town while US 1 curves northeast as the Rising Sun Bypass. US 1 intersects MD 276 (Jacob Tome Memorial Highway) and crosses Stone Run twice before turning north to the Pennsylvania state line. The highway continues across the state line toward Oxford, Pennsylvania, becoming a four-lane divided highway just across the state line.

History

Colonial and turnpike eras

The original predecessors of US 1 were a collection of dirt roads cut through the forests and farmland of central and northern Maryland in the 18th century. Construction of these roads was governed by a 1704 Act of the Province of Maryland requiring counties to oversee construction build 20-foot (6.1 m) wide roads to benefit transport of carts between population centers. The first segment of the road between Baltimore and Washington was built in 1741 between Baltimore and Elkridge as a southward extension of a road between Baltimore and Hanover, Pennsylvania. Passage across the Patapsco River at Elkridge was provided by Norwood's Ferry. In 1749, the road was blazed to Georgetown via Waterloo, Laurel, and Bladensburg. The highway north of Baltimore was a road constructed in the second half of the 18th century to connect the port of Baltimore with farms in Baltimore and Harford counties in Maryland and in Lancaster and Chester counties in Pennsylvania. From the docks in Baltimore, the road passed through Perry Hall and Kingsville on its way to Bel Air. The highway headed east from Bel Air to Churchville, then north to a crossing of the Susquehanna River at Conowingo. The road continued east to Rising Sun, then turned north toward Oxford, Pennsylvania, where the highway connected with roads to Philadelphia.

In the early 19th century, many of these dirt roads were reconstructed as turnpikes. The Washington Turnpike was chartered in 1796, but no road building ever occurred. It was not until the Washington and Baltimore Turnpike was chartered in 1812 that construction began along a 60-foot (18 m) right of way between the corner of Pratt and Eutaw Streets, then at the western city limit of Baltimore, southwest along the old dirt road to the District of Columbia boundary southwest of Bladensburg.[8] In 1817, a timber toll bridge replaced Norwood's Ferry in Elkridge. The Washington and Baltimore Turnpike had its chartered revoked in 1865 and the highway was turned over to the counties. The toll bridge at Elkridge was sold to the counties in 1869.

By 1825, a turnpike existed from Baltimore toward Bel Air. The Bel Air Turnpike was constructed from the Baltimore city line at the corner of North Avenue and Gay Street through Perry Hall and Kingsville to a junction with the Harford Turnpike at Benson in Harford County. From Benson, the Bel Air Turnpike of Harford County continued east to Main Street in Bel Air. North of Bel Air, what is now US 1 followed county highways to the Conowingo Bridge, which was first constructed around 1820.

Construction of State Road No. 1

By the beginning of the 20th century, county maintenance of the corridor that would become US 1 was becoming inadequate for the increasing amount of traffic using the roads and becoming unacceptable to advocates of better roads. As a result, the Maryland General Assembly passed the first state-aid road construction law in 1904, providing matching funds from the state for the counties to surface their major highways. As the most important corridor in the state, much of the state-aid money in the relevant counties went to the road from Baltimore to Washington. Feeling the work done by the counties with state funds was inadequate, in 1906 the Maryland General Assembly appropriated a total of $90000—$30000 each in 1906, 1907, and 1908—toward reconstruction of the Baltimore–Washington Boulevard. The 30-mile (48 km) road between the city limits of Washington and Baltimore was officially designated State Route No. 1.[9] [10] Another $174000 was appropriated by the Maryland General Assembly in 1908. The 1908 act contained stipulations requiring a realignment of the road between Beltsville and Contee to eliminate two grade crossings of the B&O Railroad as well as grade separations with the railroad at Elkridge and Winans.

By 1910, the Baltimore–Washington Boulevard was paved from the District of Columbia boundary southwest of Bladensburg north to Beltsville, with small gaps at Northeast Branch south of Hyattsville and at Paint Branch in College Park; from Contee to Elkridge; and from Hammonds Ferry Road in Halethorpe to the Baltimore city line at Gwynns Falls. The Maryland General Assembly appropriated another $120000 in 1910 for the newly-formed Maryland State Roads Commission to complete the Baltimore–Washington Boulevard. By 1915, State Route No. 1 was completely paved with the addition of the highway on a new alignment between Beltsville and Contee, the filling of the gaps in College Park and Hyattsville, and construction between Elkridge and Halethorpe. The road was constructed a minimum of 14 feet (4.3 m) in width with macadam, gravel, and concrete, with curves reduced and grades straightened. Concrete sections in Laurel, College Park, and Bladensburg were among the first concrete roads in Maryland when they were poured in 1912. Concrete girder bridges were built at Northeast Branch and the Anacostia River in Bladensburg. A concrete bridge was built over the Patapsco River at Elkridge to replace an iron bridge. Due to heavy traffic and inadequate initial construction, many of the sections constructed in the first few years of state aid had to be rebuilt.

Construction north of Baltimore

All of present-day US 1 north of Baltimore, with the exception of the road from east of Rising Sun to the Pennsylvania state line, was designated one of the original state roads by the Maryland State Roads Commission in 1909.[11] By 1910, the highway was paved from Bel Air to Kalmia. In 1910, the State Roads Commission purchased the Bel Air Turnpike, the Bel Air Turnpike of Harford County, and Conowingo Bridge from their private operators. SRC also paved its first section of the road from Oakwood to the crossing of Octoraro Creek that year. The road from Rising Sun to Sylmar Road was completed in 1911, along with Bel Air Road from the Baltimore city limits to Franklin Avenue. The paved section along Bel Air Road was extended northeast to Hamilton Avenue in 1912. That year also saw the first paving in Harford County from Kalmia to Deer Creek. The remainder of the highway in Harford County—from Little Gunpowder Falls to Bel Air and from Deer Creek to the Conowingo Bridge—was completed in 1913. Bel Air Road was finished in Baltimore County in 1914, and was paved from North Avenue to the Baltimore city limits, then just north of Erdman Avenue, in 1915. The highway was paved between the

Conowingo Bridge and Oakwood in 1914. The state road was constructed along the alignment of the turnpikes without digression except for a relocation at Gunpowder Falls to reduce the grades on the hills. The final sections of what was to become US 1 were paved in Cecil County between 1917 and 1921, when the gap between Octoraro Creek and Rising Sun was filled and the Sylmar Road link from east of Rising Sun to the Pennsylvania state line was paved.

Reconstruction of State Road No. 1

The barrage of heavy military vehicles that travelled the Baltimore–Washington Boulevard during World War I, as well as the severe winter of 1917-18, devastated the highway. As a result, in 1918 and 1919 the highway was reconstructed from end to end. Some sections of the highway were completely rebuilt with concrete, while along 18 miles (29 km) of the road concrete shoulders, among the first applied in Maryland, were added to both widen and strengthen the road. By 1919, the entire length of the Baltimore–Washington Boulevard was a minimum width of 20 feet (6.1 m). The only significant realignment during this reconstruction was the straightening out of Dead Man's Curve 0.5 miles (0.80 km) south of Elkridge.

Massive increases in traffic during the 1920s made the reconstruction of the late 1910s obsolete within the decade. Construction to expand the entire length of the Baltimore–Washington Boulevard with two 10-foot (3.0 m) shoulders to 40 feet (12 m) in width, allowing for four lanes, began in 1928. The final sections of the expanded highway, between the DC line and Bladensburg and a bypass of Elkridge, including the present underpass of the B&O Railroad, were completed in 1931. The old road through Elkridge was designated MD 477.

Reconstruction north of Baltimore

When US 1 was designated in 1927, its route through Baltimore was the same as it is today except the federal highway entered the city from the southwest along Washington Boulevard. The highway turned north onto Monroe Street, which it followed to its present course north of Wilkens Avenue.

US 1's first major relocation north of Baltimore occurred at the Susquehanna River. The Conowingo Bridge was removed before it could be inundated by the Conowingo Reservoir set to fill upon the completion of Conowingo Dam in 1928. The old approaches, which consisted of Smith Road from near Darlington to Glen Cove on the west side of the river and Connelly Road, Ragan Road, and Old Conowingo Road to the mouth of Conowingo Creek on the east side of the river, were replaced with the present highway between Smith Road and Connelly Road. Smith Road was redesignated MD 162, while the eastern approach became a western extension of MD 273.

US 1 saw three realignments north of Baltimore in 1934. A 3-mile (4.8 km) relocation near Kalmia included a new bridge over Deer Creek. The old highway, which featured 26 curves, was designated MD 590. Another relocation occurred between Conowingo and Rising Sun, featuring a new bridge over Octoraro Creek. The old highway was designated MD 591. East of Rising Sun, a 90-degree turn was bypassed at the intersection of Telegraph Road (now MD 273) and Sylmar Road by the construction of a sweeping curve to the northwest of the intersection. The bypassed portion of Sylmar Road was designated MD 592.

Bel Air Road was expanded started in 1933. US 1 was widened to 40 feet (12 m) from the Baltimore city line northeast to Joppa Road. The highway was widened to 30 feet (9.1 m) from Joppa Road to Bel Air. The 30-foot (9.1 m) road was the first construction in Maryland of a three-lane road with center turn lane.

Baltimore–Washington Boulevard receives relief

Despite the widening of US 1 between Baltimore and Washington in the late 1920s, the highway continued to be a major problem for both local and through traffic. The very high traffic levels made the highway a very attractive location for businesses serving travelers, leading to the sobriquet of "hot dog highway." The captive audience also led to the construction of over 1000 billboards between the two cities, spurring another nickname: "billboard boulevard." US 1 was also nicknamed "bloody Mary" due to the very high accident rate on the highway. After 1930, businesses were packed up to the edge of the four-lane, shoulder-less road that was largely built on an alignment poorly suited for the increasing speeds of vehicles. Thus, further expansion of the highway or safety improvements were impossible without expensive condemnation proceedings or relocating the highway. Businesses were adamantly against the latter solution.

In 1939, US 1's departure north from Washington Boulevard in Baltimore was moved south from Monroe Street to Caton Avenue just north of the city line. From Caton Avenue, US 1 followed Wilkens Avenue to Monroe Street. Caton Avenue and Wilkens Avenue were expanded between 1936 and 1938, including removal of streetcar tracks and expansion to a divided boulevard on Wilkens Avenue, to handle the increased traffic as a "through-street." By 1946, a Bypass US 1 was signed following Caton Avenue north from Wilkens Avenue, continuing on Hilton Street to North Avenue, then taking North Avenue east to rejoin US 1 at Monroe Street. Southwestern Boulevard was completed in 1950 as a dual highway through Arbutus, with grade separations with Sulphur Spring Road and Francis Avenue. US 1 was moved to its present alignment along that boulevard and Wilkens Avenue and old US 1 through Halethorpe was designated US 1 Alternate.

During World War II, US 1 was relocated from Baltimore Avenue and Bladensburg Road between Washington and Hyattsville to Rhode Island Avenue, which was widened to 36 feet (11 m) in 1940. Rhode Island Avenue between Mount Rainier and Hyattsville had originally been MD 411. The old US 1 south of Hyattsville became US 1 Alternate. Despite these relocations and upgrades, true relief did not come until the first limited-access highway between Baltimore and Washington, the Baltimore–Washington Parkway, was completed in 1954 to remove long distance traffic from US 1. With the completion of that highway and later the Harbor Tunnel Thruway (I-895) and I-95, US 1 between the two cities became a highway mainly for local traffic.

Recent improvements

Dead Man's Curve was bypassed south of Elkridge in 1946. A bypass of Laurel was constructed around 1950. In 1952, northbound US 1 was placed on Fulton Avenue in West Baltimore opposite southbound US 1 on Monroe Street. North Avenue was widened to a divided highway for much of its US 1-designated length by 1957. By 1960, this bypass had become the northbound lanes of the present one-way pair. US 1 north of Baltimore was reconstructed to modify curves and widen the road in the 1950s. The highway was reconstructed in Cecil County from Conowingo Dam to Rising Sun in 1952 and 1953 and from Rising Sun to Sylmar in 1954. US 1 in Harford County was reconstructed from Little Gunpowder Falls to north of Deer Creek in 1952 and 1953 and to Conowingo Dam between 1954 and 1956. In Howard County, the highway was resurfaced from Elkridge to Waterloo in 1954 and from Waterloo to the Patuxent River in 1956. By 1958, US 1 was being widened from Laurel to Beltsville. The highway's bypass of Rising Sun was completed in 1957. MD 273 was extended west from Sylmar through Rising Sun along the old alignment to the southern end of the bypass. The Bel Air bypass was placed under construction in 1964 and completed in 1965, resulting in the designation of US 1 Business along the old road through Bel Air. The Bel Air bypass was extended north around Hickory in 2000, with US 1 Business extended to meet the northern end of the bypass.

Junction list

County	Location	Mile [1] [2] [3] [4] [5] [6]	Destinations	Notes
Prince George's	Mount Rainier	0.00	US 1 south (Rhode Island Avenue) – Washington Eastern Avenue	District of Columbia boundary
		0.16	34th Street / Perry Street	Roundabout; 34th Street is old alignment of MD 208
	Brentwood	0.48	MD 208 (38th Street) – Colmar Manor, Cottage City, Hyattsville	
	Hyattsville	1.82	US 1 Alt. south (Baltimore Avenue) – Bladensburg	
	Riverdale Park	2.61	MD 410 (East–West Highway) – New Carrollton, Takoma Park	
	University Park	3.20	Queens Chapel Road south / Pineway west / Amherst Road east	Queens Chapel Road is former MD 500; no entry to Queens Chapel Road except buses
	College Park	3.83	College Avenue east / Regents Drive north – University of Maryland, College Park	College Avenue is former MD 203
		4.32	Paint Branch Parkway east / Campus Drive west – University of Maryland, College Park	
		5.03	MD 430 east (Greenbelt Road) to MD 193 east – Greenbelt	MD 430 is unsigned
		5.24	MD 193 west (University Boulevard) – University of Maryland, College Park, Langley Park	Partial cloverleaf interchange
		6.43	I-95 / I-495 (Capital Beltway) – Baltimore, Silver Spring, Andrews AFB	I-95/I-495 Exit 25; partial cloverleaf interchange
	Beltsville	8.00	MD 212 (Powder Mill Road) – Calverton	
	Laurel		MD 200 (Intercounty Connector)	Construction expected to begin in January 2012; interchange expected to open in 2014
		13.14	US 1 splits into one-way pair: Second Street northbound and Washington Boulevard southbound	
		13.40	Bowie Road	Old alignment of MD 197
		13.49	MD 198 east (Gorman Avenue/Fort Meade Road) – Fort Meade	
		13.58	MD 198 west (Talbott Avenue) – Burtonsville	
		13.99	Main Street	Main Street between the one-way pair is unsigned MD 979
		14.12	Bridge over Patuxent River	

County	Location	Mile	Destinations	Notes
Howard	North Laurel	14.64	US 1 rejoins from one-way pair: Washington Boulevard southbound and Second Street northbound	
		15.09	Whiskey Bottom Road – Maryland City	
	Savage	16.97	[32] MD 32 (Patuxent Freeway) – Columbia, Fort Meade	MD 32 Exit 12; cloverleaf interchange
		17.38	Guilford Road – Annapolis Junction	Former MD 732
	Jessup	19.72	[175] MD 175 (Waterloo Road) – Columbia, Fort Meade	
	Dorsey	21.17	[103] MD 103 (Meadowridge Road)	
		21.68	[100] MD 100 – Ellicott City, Glen Burnie	MD 100 Exit 6; partial cloverleaf interchange
	Elkridge	23.28	Old Washington Road north	Former MD 477; old alignment of US 1
		23.73	Montgomery Road	Old alignment of MD 103; Montgomery Road eastbound is unsigned MD 103A
		24.65	Old Washington Road south	Former MD 477; old alignment of US 1
		25.01	[895] I-895 north (Harbor Tunnel Thruway)	I-895 Exit 1; northbound exit and southbound entrance
		25.13	Bridge over Patapsco River	
Baltimore County	Arbutus	25.44	[195] I-195 (Metropolitan Boulevard) – BWI Airport, Catonsville	I-195 Exit 3
		26.02	[ALT] [1] [695] US 1 Alt. north (Washington Boulevard) to I-695 – Halethorpe	Northbound exit and southbound entrance

Baltimore City		29.18	372 MD 372 west (Wilkens Avenue)	US 1 turns onto Wilkens Avenue
		29.77	1 95 US 1 Alt. south (Caton Avenue) to I-95 / Caton Avenue north	
		31.24	Monroe Street	US 1 splits into one-way pair: Monroe Street is southbound US 1
		31.34	Fulton Avenue	US 1 northbound turns north onto Fulton Avenue
		31.52	Pratt Street	
		31.59	144 MD 144 west (Lombard Street)	
		32.06	Mulberry Street to 40 US 40 east	
		32.13	Franklin Street to 40 US 40 west	
		33.24	140 MD 140 north (Fulton Avenue) / North Avenue west	US 1 rejoins from one-way pair; US 1 turns east onto North Avenue
		33.67	129 MD 129 south (Druid Hill Avenue)	
		33.74	129 MD 129 north (McCulloh Street)	
		34.39	83 I-83 (Jones Falls Expressway) / Mount Royal Avenue south	I-83 Exit 6
		34.85	139 MD 139 north (Charles Street)	
		34.94	2 MD 2 south (St. Paul Street)	
		35.00	Calvert Street north	
		35.23	45 MD 45 north (Greenmount Avenue) / Greenmount Avenue south	
		35.79	147 MD 147 north (Harford Road) / Harford Avenue south	
		36.61	Gay Street south / North Avenue east	US 1 turns north onto Bel Air Road
		37.46	151 MD 151 (Erdman Avenue) / Erdman Avenue north	
Baltimore	Overlea	42.13	695 I-695 (Baltimore Beltway) – Towson, Essex	I-695 Exit 32; cloverleaf interchange
		43.11	43 MD 43 (White Marsh Boulevard) – White Marsh	Right-in/right-out interchange with southbound US 1
		43.37	43 MD 43 (White Marsh Boulevard) – White Marsh / Dunfield Road west	Northbound US 1 access to MD 43
	Perry Hall	44.82	Joppa Road west / Ebenezer Road east	Joppa Road is former MD 148
		51.65	Bridge over Little Gunpowder Falls	

County	Location			Notes
Harford	Benson	53.00	152 MD 152 (Mountain Road) – Fallston, Joppa	
		54.32	147 US 1 Bus. north (Bel Air Road) / MD 147 west (Harford Road) – Bel Air, Parkville	
	Bel Air North	56.26	24 MD 24 south (Vietnam Veterans Memorial Highway) – Edgewood	South end of concurrency with MD 24
		57.80	24 924 MD 24 north (Rock Spring Avenue) / MD 924 south – Bel Air, Forest Hill	North end of concurrency with MD 24; partial cloverleaf interchange
		58.90	US 1 Bus. (Conowingo Road) – Bel Air	
	Hickory	59.17	23 MD 23 west – Forest Hill	
		59.75	543 MD 543 (Ady Road) – Churchville, Pylesville	
		60.26	US 1 Bus. south (Conowingo Road) – Hickory	
	Kalmia	62.35	Forge Hill Road north	Old alignment of US 1
		65.28	Forge Hill Road south	
	Dublin	65.79	136 MD 136 (Priestford Road/Whiteford Road) – Churchville, Whiteford	
		67.85	440 MD 440 (Dublin Road)	
	Darlington	68.48	Smith Road east	Former MD 162; old alignment of US 1
		69.49	161 MD 161 south (Main Street) – Havre de Grace	
		69.67	623 MD 623 (Castleton Road) – Castleton	
		71.12	Shuresville Road	Former MD 160
		71.63	Conowingo Dam – Susquehanna River	
Cecil	Conowingo	72.08	222 MD 222 south (Susquehanna River Road) – Port Deposit, Perryville	South end of concurrency with MD 222
		73.06	222 US 222 north (Rock Springs Road) / Rowlandsville Road south – Oakwood, Lancaster	North end of concurrency with MD 222, south end of concurrency with MD 222 Truck, Rowlandsville Road is former MD 338
		74.67	591 MD 591 north (Colora Road) / Connelly Road north	Officially MD 591A, which is old alignment of US 1; Connelly Road is former MD 273 and old alignment of US 1
		75.73	591 MD 591 south (Porters Bridge Road)	Officially MD 591B; old alignment of US 1
	Rising Sun	77.04	273 MD 273 east (Rising Sun Road) – Newark, DE	Old alignment of US 1
		77.73	276 222 MD 276 / MD 222 Truck south (Jacob Tome Memorial Highway) – Port Deposit	North end of concurrency with MD 222 Truck
		80.86	1 US 1 north – Oxford, PA, Philadelphia	Pennsylvania state line
1.000 mi = 1.609 km; 1.000 km = 0.621 mi				

References

[1] "Highway Location Reference: Prince George's County" (http://www.marylandroads.com/Location/2009_PRINCEGEORGES.pdf)
     (PDF). Maryland State Highway Administration. 2009. . Retrieved 2010-09-21.

[2] "Highway Location Reference: Howard County" (http://www.marylandroads.com/Location/2009_HOWARD.pdf) (PDF). Maryland
     State Highway Administration. 2009. . Retrieved 2010-09-21.

[3] "Highway Location Reference: Baltimore County" (http://www.marylandroads.com/Location/2009_BALTIMORE.pdf) (PDF). Maryland
     State Highway Administration. 2009. . Retrieved 2010-09-21.

[4] "Highway Location Reference: Baltimore City" (http://www.marylandroads.com/Location/2005_BALTIMORECITY.pdf) (PDF).
     Maryland State Highway Administration. 2005. . Retrieved 2010-09-21.

[5] "Highway Location Reference: Harford County" (http://www.marylandroads.com/Location/2009_HARFORD.pdf) (PDF). Maryland
     State Highway Administration. 2009. . Retrieved 2010-09-21.

[6] "Highway Location Reference: Cecil County" (http://www.marylandroads.com/Location/2009_CECIL.pdf) (PDF). Maryland State
     Highway Administration. 2009. . Retrieved 2010-09-21.

[7] Weingroff, Richard F. (January 9, 2009). "From Names to Numbers: The Origins of the U.S. Numbered Highway System" (http://wwwcf.
     fhwa.dot.gov/infrastructure/numbers.cfm). Federal Highway Administration. . Retrieved April 21, 2009.

[8] Prince George's County Circuit Court, Land Survey, Subdivision, and Condominium Plats, Plat Drawer 11, File 17, Turnpike Road from the
     District of Columbia to the City of Baltimore (http://plato.mdarchives.state.md.us/msa/stagser/s1500/s1529/cfm/dsp_unit.
     cfm?county=pg&qualifier=C&series=2482&unit=254)

[9] (http://www.mdarchives.state.md.us/megafile/msa/speccol/sc2900/sc2908/000001/000377/html/am377--767.html)

[10] (http://www.mdroads.com/routes/001-006.html#us001)

[11] "Crossroads: The History of Perry Hall, Maryland," by David Marks, pp. 75-77; 89-90. Published by Gateway Press, Inc., 1999.

External links

- US 1 @ MDRoads.com (http://www.mdroads.com/routes/us001.html)
- United States Numbered Highways, 1989 Edition (http://cms.transportation.org/?siteid=68&pageid=1760), by
  the AASHTO

Pennsylvania_Route_24

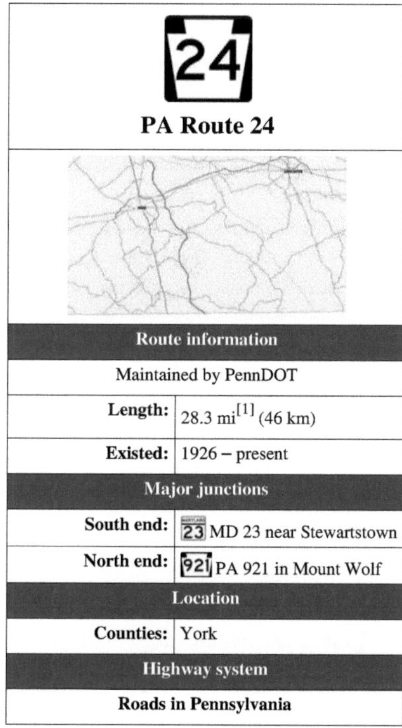

Route information	
Maintained by PennDOT	
Length:	28.3 mi[1] (46 km)
Existed:	1926 – present
Major junctions	
South end:	MD 23 near Stewartstown
North end:	PA 921 in Mount Wolf
Location	
Counties:	York
Highway system	
Roads in Pennsylvania	

Pennsylvania Route 24 (PA 24) is an 29-mile (47 km) long state highway located within central Pennsylvania in the York County area. Its southern terminus is at the Mason–Dixon Line near Stewartstown, where PA 24 continues into Maryland as Maryland Route 23. The northern terminus is at Pennsylvania Route 921 in Mount Wolf.

Route description

PA 24 begins at the Maryland border in Hopewell Township, heading to the northwest on two-lane undivided Barrens Road South. The road continues into Maryland as MD 23. From the state line, the route passes through agricultural areas with scattered residences. PA 24 enters Stewartstown and becomes Main Street, heading north past homes. In the center of town, the route forms a short concurrency with PA 851. The road passes more residences and a few businesses before leaving Stewartstown for Hopewell Township again, becoming Barrens Road North. PA 24 continues through more open farmland with occasional patches of woods and homes. Upon passing through the community of Rinely, the route crosses into North Hopewell Township and continues northwest through more rural areas on Winterstown Road. PA 24 enters Winterstown and becomes Main Street, passing a few homes as it intersects PA 216. After this intersection, the road passes more farmland and heads back into North Hopewell Township, with the road name changing back to Winterstown Road. The route heads through more farm areas with some houses, curving to the northeast as it becomes the border between York Township to the west and Windsor Township to the east.[1] [2]

The road passes homes as it continues into Red Lion, where it turns north onto Main Street. PA 24 is lined with more residences as it heads northwest through Red Lion and intersects PA 74 in the center of town. Past this intersection, the route continues past a mix of homes and businesses in the downtown area, coming to the PA 624 junction. The road passes more residences before leaving Red Lion and becoming the York Township-Windsor Township border again, with the road name becoming Cape Horn Road. PA 24 passes a mix of suburban homes and businesses, running through East Yoe. The route turns more to the north as it continues through a mix of farmland and residential development. The road makes a turn northwest into suburban neighborhoods and turns north again as it comes to an intersection with PA 124 in a commercial area. Past PA 124, PA 24 continues north on Edgewood Road and enters Springettsbury Township, passing wooded residential neighborhoods. The route widens into a four-lane divided highway as it reaches the PA 462 junction.[1] [2]

From here, the road becomes Mount Zion Road and passes homes before running near industrial establishments, coming to a bridge over a Norfolk Southern railroad line. After an interchange with US 30, PA 24 enters commercial areas and passes to the east of the York Galleria shopping mall. Past the mall, the route narrows back into a two-lane undivided road and heads northwest through a mix of wooded residential areas and some fields, passing through Mount Zion. PA 24 makes a turn north onto an unnamed road and crosses the Codorus Creek into Manchester Township, passing a mix of residential neighborhoods and farmland. The route passes through Starview and turns to the northwest, heading into Mount Wolf. Here, PA 24 becomes Center Street and passes homes before turning southwest onto Main Street and ending at the eastern terminus of PA 921 a short distance later.[1] [2]

Major intersections

The entire route is in York County.

Location	Mile	Destinations	Notes
Stewartstown	0.0	MD 23 south (Norrisville Road)	Western terminus of MD 23
	3.6	PA 851 east (College Avenue)	South end of concurrency
	3.7	PA 851 west (Pennsylvania Avenue)	North end of concurrency
Winterstown	9.8	PA 216	Eastern terminus of PA 216
Red Lion	15.0	PA 74 (Broadway)	
	15.2	PA 624 (High Street)	Southern terminus of PA 624
East York	20.3	PA 124	
	21.6	PA 462 (Market Street)	
	22.2	US 30	Interchange
Mount Wolf	28.8	PA 921 (Main Street)	Eastern terminus of PA 921
1.000 mi = 1.609 km; 1.000 km = 0.621 mi			

See also

Spur routes

- Pennsylvania Route 124
- Pennsylvania Route 624

References

[1] Google, Inc. *Google Maps – overview of Pennsylvania Route 24* (http://maps.google.com/maps?f=d&source=s_d&saddr=MD+23+and+
     PA+24&daddr=39.9165415,-76.6301181+to:PA+24+and+PA+921&
     geocode=FTcgXgIdxuJv-ykVnCniK3rIiTGcSa0_o5xF6w;Ff0TYQIdmrdu-ylV6XXWIobIiTEU60neb63d1A;FRZSYwIdYoVt-ykN1F46opHIiTEllL3oOnbaTQ
     hl=en&mra=ls&sll=39.836799,-76.613653&sspn=0.00748,0.021136&ie=UTF8&ll=39.896041,-76.643372&spn=0.47412,1.
     352692&t=h&z=10&via=1) (Map). Cartography by Google, Inc. . Retrieved February 25, 2011.
[2] PennDOT (2011). *York County, Pennsylvania Highway Map* (ftp://ftp.dot.state.pa.us/public/pdf/BPR_pdf_files/Maps/GHS/
     Roadnames/york_GHSN.PDF) (Map). . Retrieved February 22, 2011.

Maryland_Route_165

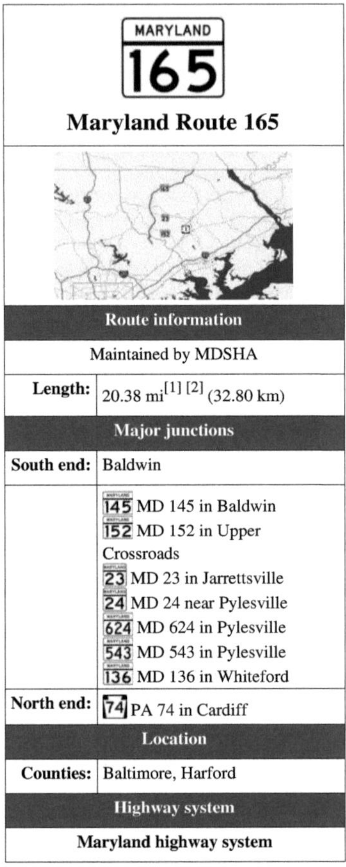

Maryland Route 165

Route information
Maintained by MDSHA

Length:	20.38 mi[1] [2] (32.80 km)

Major junctions	
South end:	Baldwin
	MD 145 in Baldwin MD 152 in Upper Crossroads MD 23 in Jarrettsville MD 24 near Pylesville MD 624 in Pylesville MD 543 in Pylesville MD 136 in Whiteford
North end:	PA 74 in Cardiff

Location	
Counties:	Baltimore, Harford

Highway system
Maryland highway system

Maryland Route 165 (**MD 165**) is a state highway in the U.S. state of Maryland. The state highway runs 20.38 miles (32.80 km) from Baldwin north to the Pennsylvania state line in Cardiff, where the highway continues as Pennsylvania Route 74 (PA 74). MD 165 passes through western and northern Harford County, where it connects the communities of Jarrettsville, Pylesville, and Whiteford. The state highway was constructed as part of MD 24 through Pylesville and Whiteford in the late 1910s and early 1920s. MD 165 from Baldwin through Jarrettsville to west of Pylesville was built in the late 1920s and early 1930s. When MD 24 was rerouted in 1933, MD 165 was extended along that highway's old routing through Pylesville and Whiteford, much of which was relocated in 1960.

Route description

MD 165 begins at a seemingly arbitrary location along Baldwin Mill Road; this location was once where the highway intersected the Maryland and Pennsylvania Railroad. Baldwin Mill Road continues south as part of a series of county roads—Fork Road, Sunshine Avenue, and Bradshaw Road—that parallel Little Gunpowder Falls along the eastern edge of Baltimore County through the communities of Baldwin, Fork, Kingsville, and Bradshaw. MD 165 heads north as a two-lane road that meets the eastern end of MD 145 (Sweet Air Road) before curving east to cross Little Gunpowder Falls into Harford County.[1] [3] The state highway passes through a mix of farms and forests on its way to Upper Crossroads, where the state highway intersects MD 152 (Fallston Road). MD 165 crosses the West Branch of Winters Run, passes through the hamlet of Putnam, and intersects the western end of East–West Highway, which carries MD 23, just south of Jarrettsville. The two state highways run concurrently to the center of the village, where MD 23 turns west onto Norrisville Road at the intersection with its old alignment, Jarrettsville Road.[2] [3]

MD 165 continues north from Jarrettsville as Federal Hill Road, which passes through the namesake hamlet then curves east and crosses Deer Creek. At Bush's Corner, where the state highway meets MD 24 (Rocks Road) at a roundabout, its name changes to Pylesville Road and the highway passes a trio of schools. North Harford High School on the south side of the highway is connected to the middle and elementary schools on the north side by a pedestrian tunnel. MD 165 continues east past the southern terminus of MD 624 (Graceton Road) to Pylesville, where the highway curves to the northeast and meets the northern end of MD 543 (Ady Road) and short distance north of the village of Street. The state highway crosses Broad Creek and parallels Old Pylesville Road northeast through the village of Whiteford. Both the old road and modern MD 165 intersect MD 136 (Whiteford Road). Old Pylesville Road passes through Cardiff, which contains the Slate Ridge School within the Whiteford-Cardiff Historic District preserving the area's slate heritage. MD 165 bypasses the village on the way to its northern terminus at the Pennsylvania state line. The highway continues north as PA 74 (Delta Road), which bypasses Cardiff's neighbor across the state line, Delta.[2] [3]

History

The first portion of MD 165 to be paved was constructed as part of MD 24, which originally followed Pylesville Road to Cardiff instead of continuing north toward Fawn Grove, Pennsylvania.[4] The portion of Pylesville Road between Graceton Road and Broad Creek was paved by 1910.[5] Pylesville Road from Graceton Road west to Bush's Corner was constructed as a 15-foot (4.6 m) wide concrete road by 1919.[6] [7] The road from Broad Creek north to Cardiff was constructed as a macadam road between 1921 and 1923.[7] [8] Pylesville Road was expanded to a width of 20 feet (6.1 m) from Pylesville to Cardiff between 1926 and 1930.[9] [10] MD 24 and MD 165 were moved to their present corridors north of Bush's Corner in 1933.[11]

The portion of MD 165 from its southern terminus at the Maryland and Pennsylvania Railroad north to the MD 145 intersection was built as a concrete road in 1926 and 1927 as part of the highway between Cockeysville and Baldwin. MD 165 was constructed north from Jarrettsville starting in 1924; the concrete highway reached Federal Hill in 1927.[4] [9] A short section of the state highway south of Jarrettsville was started in 1926 and completed in 1928.[9] [12] MD 165 was extended north from Federal Hill to near Deer Creek around 1930.[13] The state highway was completed between Jarrettsville and Baldwin in 1932. MD 165 was completed when the final section of the modern highway was finished from west of Deer Creek to Bush's Corner in 1933.[11] [14] MD 165 was relocated from Pylesville to the Pennsylvania state line around 1960, leaving behind old Pylesville Road.[15] The roundabout at the MD 24 junction was built in 2000.[16]

Junction list

County	Location	Mile [1] [2]	Destinations	Notes
Baltimore	Baldwin	0.00	Baldwin Mill Road south – Fork	Southern terminus
		0.83	**145** MD 145 west (Sweet Air Road) – Jacksonville	
Harford	Upper Crossroads	3.59	**152** MD 152 (Fallston Road) – Fallston	
	Jarrettsville	6.92	**23** MD 23 east (East–West Highway) – Forest Hill	South end of concurrency with MD 23
		8.20	**23** MD 23 north (Norrisville Road) – Norrisville	North end of concurrency with MD 23
	Pylesville	15.13	**24** MD 24 (Rocks Road) – Forest Hill, Fawn Grove	Roundabout
		16.34	**624** MD 624 north (Graceton Road) – Graceton	
		17.15	**543** MD 543 south (Ady Road) – Hickory	
	Whiteford	19.63	**136** MD 136 (Whiteford Road) – Graceton, Dublin	
	Cardiff	20.38	**74** PA 74 north (Delta Road) – Delta	Northern terminus; Pennsylvania state line
1.000 mi = 1.609 km; 1.000 km = 0.621 mi				

References

[1] "Highway Location Reference: Baltimore County" (http://www.roads.maryland.gov/Location/2010_BALTIMORE.pdf) (PDF). Maryland State Highway Administration. 2010. . Retrieved 2011-07-19.

[2] "Highway Location Reference: Harford County" (http://www.roads.maryland.gov/Location/2010_HARFORD.pdf) (PDF). Maryland State Highway Administration. 2010. . Retrieved 2011-07-19.

[3] Google, Inc. *Google Maps – Maryland Route 165* (http://maps.google.com/maps?saddr=E+North+Ave&daddr=Harford+Rd+ to:MD-147+N/Harford+Rd&hl=en&ll=39.410733,-76.490936&spn=0.250931,0.617294&sll=39.5051,-76.384678&sspn=0. 015662,0.038581&geocode=FVfaVwIdpy9v-w;FeRZWQIdJYJw-w;FYrXWgIdVoJy-w&mra=ls&t=h&z=11) (Map). Cartography by Google, Inc. . Retrieved 2011-07-19.

[4] Maryland Geological Survey. *Map of Maryland: Showing State Road System and State Aid Roads* (Map) (1927 ed.).

[5] Maryland Geological Survey. *Map of Maryland* (Map) (1910 ed.).

[6] *Report of the State Roads Commission of Maryland* (http://www.archive.org/details/annualreportsofs1916mary) (1916–1919 ed.). Baltimore: Maryland State Roads Commission. 1920-01. p. 39. . Retrieved 2011-07-21.

[7] Maryland Geological Survey. *Map of Maryland: Showing State Road System and State Aid Roads* (Map) (1921 ed.).

[8] Maryland Geological Survey. *Map of Maryland: Showing State Road System and State Aid Roads* (Map) (1923 ed.).

[9] *Report of the State Roads Commission of Maryland* (http://www.archive.org/details/annualreportsofs1924mary) (1924–1926 ed.). Baltimore: Maryland State Roads Commission. 1927-01. pp. 43, 45, 69, 84. . Retrieved 2011-07-21.

[10] *Report of the State Roads Commission of Maryland* (http://www.archive.org/details/reportofstateroa1927mary) (1927–1930 ed.). Baltimore: Maryland State Roads Commission. 1930-10-01. pp. 82, 214. . Retrieved 2011-07-21.

[11] Maryland Geological Survey. *Map of Maryland Showing State Road System: State Aid Roads and Improved County Road Connections* (Map) (1933 ed.).

[12] Maryland Geological Survey. *Map of Maryland: Showing State Road System and State Aid Roads* (Map) (1928 ed.).

[13] Maryland Geological Survey. *Map of Maryland Showing State Road System: State Aid Roads and Improved County Road Connections* (Map) (1930 ed.).

[14] *Report of the State Roads Commission of Maryland* (http://www.archive.org/details/reportofstateroa1931mary) (1931–1934 ed.). Baltimore: Maryland State Roads Commission. 1934-12-28. pp. 320, 338–339. . Retrieved 2011-07-21.

[15] Maryland State Roads Commission. *Maryland: Official Highway Map* (Map) (1960 ed.).

[16] "Highway Location Reference: Harford County" (http://www.roads.maryland.gov/Location/2000_HARFORD.pdf) (PDF). Maryland State Highway Administration. 2000. . Retrieved 2011-07-21.

External links

- MDRoads: MD 165 (http://www.mdroads.com/routes/160-179.html#md165)

Maryland_Route_146

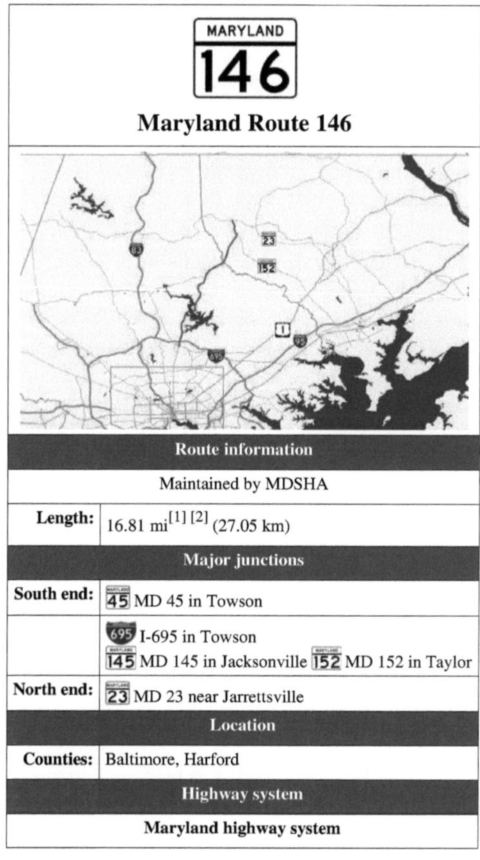

Maryland Route 146	
Route information	
Maintained by MDSHA	
Length:	16.81 mi[1] [2] (27.05 km)
Major junctions	
South end:	MD 45 in Towson
	I-695 in Towson MD 145 in Jacksonville MD 152 in Taylor
North end:	MD 23 near Jarrettsville
Location	
Counties:	Baltimore, Harford
Highway system	
Maryland highway system	

Maryland Route 146 (**MD 146**) is a state highway in the U.S. state of Maryland. The state highway runs 16.81 miles (27.05 km) from MD 45 in Towson north to MD 23 near Jarrettsville. MD 146 connects Towson with Loch Raven Reservoir, an impoundment of Gunpowder Falls. The state highway also serves the northern Baltimore County community of Jacksonville and Jarrettsville in western Harford County. MD 146 was constructed as two different state highways on either side of Loch Raven Reservoir. The section of the state highway in Towson was built in the 1910s and the portion through Jacksonville to Jarrettsville was constructed in the late 1920s and early 1930s. The gap in MD 146 through Loch Raven Reservoir was filled in two steps of maintenance swaps in the early 1960s and late 1970s.

Route description

MD 146 begins at the Towson Roundabout, a five-leg, racetrack-shaped roundabout in the center of Towson. The roundabout also features MD 45 (York Road), which heads south toward Baltimore and northwest toward Lutherville, and Joppa Road, a county-maintained highway that runs west toward Brooklandville and east toward Parkville. MD 146 heads northeast as Dulaney Valley Road, a four-lane divided boulevard that passes along the west side of Towson Town Center and the campus of Goucher College. The state highway meets Interstate 695 (Baltimore Beltway) at a partial cloverleaf interchange. The ramp from the westbound Beltway to MD 146 connects via Hampton Lane, which heads east past Towson United Methodist Church toward Hampton National Historic Site. North of the Beltway, MD 146 is a five-lane road with center turn lane that passes between the suburbs of Hampton, Maryland to the east and Lutherville to the west. MD 146 reduces to two lanes just north of Seminary Road and its traversal of Long Quarter Branch, which flows into Loch Raven Reservoir. The state highway passes between Timonium to the west and the reservoir parkland on the east.[1] [3]

North of Timonium Road, MD 146 fully enters Loch Raven Reservoir Park. At Old Bosley Road, the state highway curves east and crosses the reservoir. Just east of the bridge, Dulaney Valley Road continues east while MD 146 veers north as Jarrettsville Pike. The state highway passes by the historic home Eagle's Nest as it leaves the reservoir valley. MD 146 intersects Sunnybrook Road and Merrymans Mill Road in the hamlet of Sunnybrook before passing through Jacksonville, a community also known as Phoenix that is centered at the highway's intersection with MD 145. MD 145 heads east as Sweet Air Road toward Baldwin and west as Paper Mill Road toward Cockeysville.[1] [3] North of Jacksonville, MD 146 crosses Little Gunpowder Falls into Harford County. The state highway passes through the communities of Hess and Taylor, where the highway passes the Ladew Topiary Gardens and meets the northern end of MD 152 (Fallston Road). MD 146 reaches its northern terminus at its intersection with MD 23 (Norrisville Road) in the hamlet of Madonna west of Jarrettsville. The roadway continues north as county-maintained Madonna Road.[2] [3]

History

MD 146's predecessor highways included a pair of turnpikes. The Dulaney's Valley and Towsontown Turnpike connected the Baltimore and Yorktown Turnpike at Towson with Meredith's Ford, a shallow spot in Gunpowder Falls before Loch Raven Reservoir was formed. A pair of turnpikes began east of the ford: the Dulaney's Valley and Sweet Air Turnpike east to the community of Knoebel at what is now the intersection of Dulaney Valley Road and Manor Road, and the Jarrettsville Turnpike north from the ford through Jacksonville to Little Gunpowder Falls.[4] The first section of modern MD 146 was constructed as a 14-foot (4.3 m) wide macadam road from York Road north 0.5 miles (0.80 km) in 1915.[5] This road was resurfaced in

The spire of Towson United Methodist Church is a prominent landmark at the junction of MD 146 and I-695.

concrete and extended to the southern edge of the Loch Raven Reservoir reservation just north of Seminary Road by 1921.[6]

The first portion of MD 146 north of Loch Raven Reservoir was a concrete road from the northern edge of the Loch Raven Reservoir reservation to MD 145 in Jacksonville built in 1928.[7] The state highway was extended north to Little Gunpowder Falls in 1929.[8] [9] The Harford County section of MD 146 was started in 1930 and completed in 1932.[9] [10] [11] An additional section of the state highway was built on Madonna Road from the junction with MD 23 north to Nelson Mill Road in 1939.[12] [13] MD 146 was truncated at MD 23 when the Madonna Road segment was transferred to county maintenance in 1955.[14] The section of the state highway south of Loch Raven Reservoir was originally designated MD 144 but became a disjoint segment of MD 146 by 1940.[11] [15] MD 144 was later reused for bypassed sections of U.S. Route 40 between Cumberland and Baltimore.

The southern segment of MD 146 was widened to 18 feet (5.5 m) in 1946.[16] MD 146 from the center of Towson to the Baltimore Beltway was expanded to a divided highway concurrent with the construction of the highway's cloverleaf interchange with the Beltway between 1955 and 1958.[17] [18] The portions of Dulaney Valley Road and Jarrettsville Pike through the Loch Raven Reservoir reservation were originally maintained by the Baltimore City Department of Transportation since the reservoir is owned by the city of Baltimore. These sections were transferred to state maintenance in two steps. MD 146 was extended north from near Seminary Road to just west of the bridge over the reservoir in 1963.[19] The bridge and the highway to the northern edge of the reservoir reservation were transferred to state maintenance in 1979, closing the gap between the two sections of MD 146.[20]

Junction list

County	Location	Mile [1] [2]	Destinations	Notes
Baltimore	Towson	0.00	45 MD 45 (York Road) / Joppa Road – Baltimore	Southern terminus; Towson Roundabout
		0.78	695 I-695 (Baltimore Beltway) – Pikesville, Essex	I-695 Exit 27
	Loch Raven Reservoir	4.99	Dulaney Valley Road east	MD 146 continues north as Jarrettsville Pike; Dulaney Valley Road is unsigned MD 146A
	Jacksonville	9.24	145 MD 145 (Sweet Air Road) – Cockeysville, Baldwin	
Harford	Taylor	15.13	152 MD 152 south (Fallston Road) – Fallston	
	Jarrettsville	16.81	23 MD 23 (Norrisville Road) / Madonna Road north – Norrisville, Bel Air	Northern terminus
1.000 mi = 1.609 km; 1.000 km = 0.621 mi				

Auxiliary route

MD 146A is the designation for the 0.22-mile (0.35 km) section of Dulaney Valley Road immediately east of the road's intersection with MD 146, which is just east of MD 146's bridge over Loch Raven Reservoir. Dulaney Valley Road continues east as a road maintained by the Baltimore City Department of Transportation, which maintains roads within the park reservation surrounding the reservoir.[1] [21] This section of Dulaney Valley Road was transferred from city to state maintenance in 2009.[22]

References

[1] "Highway Location Reference: Baltimore County" (http://www.roads.maryland.gov/Location/2010_BALTIMORE.pdf) (PDF). Maryland State Highway Administration. 2010. . Retrieved 2011-07-24.

[2] "Highway Location Reference: Harford County" (http://www.roads.maryland.gov/Location/2010_HARFORD.pdf) (PDF). Maryland State Highway Administration. 2010. . Retrieved 2011-07-24.

[3] Google, Inc. *Google Maps – Maryland Route 146* (http://maps.google.com/maps?saddr=Dulaney+Valley+Rd&daddr=Jarrettsville+Pike&hl=en&ll=39.508809,-76.555481&spn=0.250578,0.617294&sll=39.508809,-76.555481&sspn=0.250578,0.617294&geocode=FVo6WQIdcSVv-w;FY5zXAIdNXRw-w&mra=ls&t=h&z=11) (Map). Cartography by Google, Inc. . Retrieved 2011-07-24.

[4] Clark, William Bullock (1899). *Report on the Highways of Maryland* (http://books.google.com/books?id=b9l9AAAAIAAJ&printsec=frontcover#v=onepage&q&f=false). Baltimore: Maryland Geological Survey. p. 218. . Retrieved 2011-07-24.

[5] *Report of the State Roads Commission of Maryland* (http://www.archive.org/details/annualreportsofs1912mary) (1912–1915 ed.). Baltimore: Maryland State Roads Commission. 1916-05. p. 122. . Retrieved 2011-07-24.

[6] Maryland Geological Survey. *Map of Maryland: Showing State Road System and State Aid Roads* (Map) (1921 ed.).

[7] Maryland Geological Survey. *Map of Maryland: Showing State Road System and State Aid Roads* (Map) (1928 ed.).

[8] Maryland Geological Survey. *Map of Maryland Showing State Road System: State Aid Roads and Improved County Road Connections* (Map) (1930 ed.).

[9] *Report of the State Roads Commission of Maryland* (http://www.archive.org/details/reportofstateroa1927mary) (1927–1930 ed.). Baltimore: Maryland State Roads Commission. 1930-10-01. pp. 198, 214. . Retrieved 2011-07-24.

[10] Maryland Geological Survey. *Map of Maryland Showing State Road System: State Aid Roads and Improved County Road Connections* (Map) (1933 ed.).

[11] *Report of the State Roads Commission of Maryland* (http://www.archive.org/details/reportofstateroa1931mary) (1931–1934 ed.). Baltimore: Maryland State Roads Commission. 1934-12-28. pp. 21, 338. . Retrieved 2011-07-24.

[12] Maryland State Roads Commission. *General Highway Map: State of Maryland* (Map) (1939 ed.).

[13] *Report of the State Roads Commission of Maryland* (http://www.archive.org/details/reportofstateroa1939mary) (1939–1940 ed.). Baltimore: Maryland State Roads Commission. 1941-03-15. p. 104. . Retrieved 2011-07-24.

[14] Maryland State Roads Commission. *Maryland: Official Highway Map* (Map) (1955 ed.).

[15] Maryland State Roads Commission. *Map of Maryland Showing Highways and Points of Interest* (Map) (1940 ed.).

[16] *Report of the State Roads Commission of Maryland* (http://www.archive.org/details/reportofstateroa1945mary) (1945–1946 ed.). Baltimore: Maryland State Roads Commission. 1947-02-01. p. 98. . Retrieved 2011-07-24.

[17] *Report of the State Roads Commission of Maryland* (http://www.archive.org/details/reportofstateroa1955mary) (1955–1956 ed.). Baltimore: Maryland State Roads Commission. 1956-11-02. p. 164. . Retrieved 2011-07-24.

[18] "Major Transportation Milestones in the Baltimore Region Since 1940" (http://www.baltometro.org/reports/MajorTransMilestones.pdf) (PDF). Baltimore Metropolitan Council. 2005-12-01. p. 6. . Retrieved 2011-07-24.

[19] Maryland State Roads Commission. *Maryland: Official Highway Map* (Map) (1963 ed.).

[20] Maryland State Highway Administration. *Maryland: Official Highway Map* (Map) (1979-80 ed.).

[21] Google, Inc. *Google Maps – Maryland Route 146A* (http://maps.google.com/maps?saddr=Dulaney+Valley+Rd&daddr=Dulaney+Valley+Rd&hl=en&ll=39.464344,-76.574492&spn=0.003918,0.009645&sll=39.464344,-76.574492&sspn=0.001959,0.004823&geocode=Fc0sWgIdB4lv-w;FU0vWgIdvJhv-w&mra=ls&t=h&z=17) (Map). Cartography by Google, Inc. . Retrieved 2011-07-24.

[22] "Highway Location Reference: Baltimore County" (http://www.roads.maryland.gov/Location/2009_BALTIMORE.pdf) (PDF). Maryland State Highway Administration. 2009. . Retrieved 2011-07-24.

External links

- MDRoads: MD 146 (http://www.mdroads.com/routes/140-159.html#md146)

Maryland_Route_138

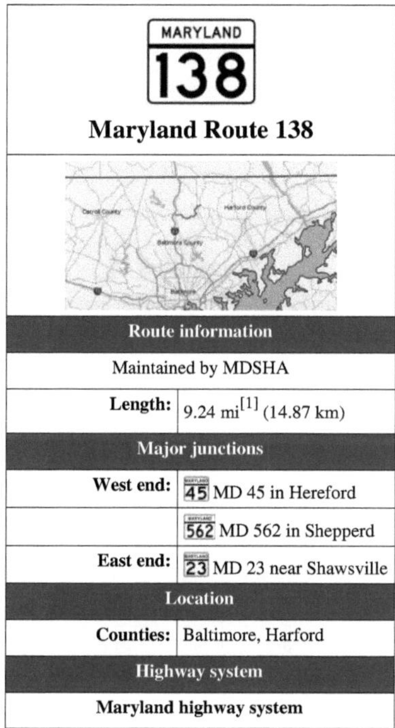

Maryland Route 138 (**MD 138**) is a state highway in the U.S. state of Maryland. The state highway runs 9.24 miles (14.87 km) from MD 45 in Hereford east to MD 23 near Shawsville. MD 138 connects northern Baltimore County with northwestern Harford County via the community of Monkton on Gunpowder Falls. The first section of the state highway was built east of Monkton in the 1910s. The remainder of MD 138 was built west of Monkton in the mid-1920s and east of Monkton in the early to mid-1930s. The state highway through Monkton was maintained by Baltimore County from the late 1960s to the mid-1990s, during which the highway was relocated at Gunpowder Falls.

Route description

MD 138 begins at an intersection with MD 45 (York Road) in Hereford, one block south of MD 45's intersection with MD 137 (Mount Carmel Road). The state highway heads east as two-lane Monkton Road to the village of Monkton, where the highway crosses Gunpowder Falls and intersects the Northern Central Railroad Trail. On the east edge of the village, MD 138 veers east onto Shepperd Road and follows Charles Run upstream out of the valley. The state highway passes through the area patented in the 18th century as My Lady's Manor. In the hamlet of Shepperd, Shepperd Road ends at Troyer Road. MD 138 turns north onto Troyer Road and Troyer Road heads south as MD 562. The state highway passes through Troyer and curves along a ridge above the headwaters of Little Gunpowder Falls. MD 138 crosses the Baltimore–Harford county line and reaches its eastern terminus at MD 23 (Norrisville Road) in the hamlet of Blackhorse near Shawsville. Troyer Road continues east on the other side of the intersection as a county highway.[1] [2]

History

The first section of pavement along modern MD 138 was built along Shepperd Road from Monkton to J.M. Pearce Road between 1915 and 1921.[3] [4] A concrete road was constructed from Hereford to just west of Gunpowder Falls between 1925 and 1927.[5] [6] The gap in the improved road through Monkton was filled in 1928 and a bridge for the Northern Central Railroad to cross over MD 138 was built shortly after 1930.[7] [8] The state highway from J.M. Pearce Road to Shepperd and along Troyer Road to the Harford County line was constructed between 1930 and 1932.[8] [9] [10] The short portion of MD 138 in Harford County was completed in 1936.[11] [12]

MD 138 from just west of Gunpowder Falls through Monkton to east of Wesley Chapel Road was transferred from state to county maintenance in 1968.[13] The county section of the highway was extended west to near Piney Hill Road in 1972.[14] Baltimore County relocated the highway around and built a new bridge over Gunpowder Falls in 1975.[15] The old highway followed what is now Old Monkton Road on either side of the river; east of the river, the highway had a right-angle turn at its railroad crossing.[16] The gap in MD 138 was reduced to the section from just west of the new bridge to the Monkton Road – Shepperd Road intersection in 1987.[17] MD 138 was united when that section was transferred to state maintenance in 1995.[18]

Junction list

County	Location	Mile [1]	Intersection	Notes
Baltimore	Hereford	0.00	45 137 MD 45 (York Road) to MD 137 (Mount Carmel Road) – Parkton, Cockeysville	Western terminus
	Shepperd	6.12	562 MD 562 south (Troyer Road)	
Harford	Shawsville	9.24	23 MD 23 (Norrisville Road) / Troyer Road east – Norrisville, Jarrettsville	Eastern terminus
1.000 mi = 1.609 km; 1.000 km = 0.621 mi				

References

[1] Maryland State Highway Administration (2010). *Highway Location Reference* (http://www.marylandroads.com/pages/hlr. aspx?PageId=832). . Retrieved 2011-07-31.

  • Baltimore County (http://www.roads.maryland.gov/Location/2010_BALTIMORE.pdf) (PDF)
  • Harford County (http://www.roads.maryland.gov/Location/2010_HARFORD.pdf) (PDF)

[2] Google, Inc. *Google Maps – Maryland Route 138* (http://maps.google.com/maps?saddr=MD-138+E/Monkton+Rd&daddr=MD-138+E/ Troyer+Rd&hl=en&sll=39.589204,-76.660216&sspn=0.007822,0.01929&geocode=FeIUXAIdzjZu-w;FayfXAIdxuxv-w&mra=ls& t=h&z=13) (Map). Cartography by Google, Inc. . Retrieved 2011-07-31.

[3] *Report of the State Roads Commission of Maryland* (http://www.archive.org/details/annualreportsofs1912mary) (1912–1915 ed.). Baltimore: Maryland State Roads Commission. 1916-05. p. 128. . Retrieved 2011-07-31.

[4] Maryland Geological Survey. *Map of Maryland: Showing State Road System and State Aid Roads* (Map) (1921 ed.).

[5] *Report of the State Roads Commission of Maryland* (http://www.archive.org/details/annualreportsofs1924mary) (1924–1926 ed.). Baltimore: Maryland State Roads Commission. 1927-01. pp. 44. 69. . Retrieved 2011-07-31.

[6] Maryland Geological Survey. *Map of Maryland: Showing State Road System and State Aid Roads* (Map) (1927 ed.).

[7] Maryland Geological Survey. *Map of Maryland: Showing State Road System and State Aid Roads* (Map) (1928 ed.).

[8] *Report of the State Roads Commission of Maryland* (http://www.archive.org/details/reportofstateroa1927mary) (1927–1930 ed.). Baltimore: Maryland State Roads Commission. 1930-10-01. pp. 68, 199. . Retrieved 2011-07-31.

[9] *Report of the State Roads Commission of Maryland* (http://www.archive.org/details/reportofstateroa1931mary) (1931–1934 ed.). Baltimore: Maryland State Roads Commission. 1934-12-28. p. 320. . Retrieved 2011-07-31.

[10] Maryland Geological Survey. *Map of Maryland Showing State Road System: State Aid Roads and Improved County Road Connections* (Map) (1933 ed.).

[11] *Report of the State Roads Commission of Maryland* (http://www.archive.org/details/reportofstateroa1935mary) (1935–1936 ed.). Baltimore: Maryland State Roads Commission. 1936-12-04. p. 81. . Retrieved 2011-07-31.

[12] Maryland State Roads Commission. *Map of Maryland Showing State Road System* (Map) (1936 ed.).

[13] Maryland State Roads Commission. *Maryland: Official Highway Map* (Map) (1968 ed.).

[14] Maryland State Highway Administration. *Maryland: Official Highway Map* (Map) (1972 ed.).

[15] "NBI Structure Number: 200000B-0014010" (http://nationalbridges.com/). *National Bridge Inventory*. . Retrieved 2011-07-31.

[16] United States Geological Survey (1974-07-01). *Monkton, Maryland, United States* (http://msrmaps.com/image.aspx?T=2&S=12& Z=18&X=451&Y=5477&W=1&qs=) (Map). Topo Map. . Retrieved 2011-07-31.

[17] Maryland State Highway Administration. *Maryland: Official Highway Map* (Map) (1987 ed.).

[18] Maryland State Highway Administration. *Maryland: Official Highway Map* (Map) (1995 ed.).

External links

- MDRoads: MD 138 (http://www.mdroads.com/routes/120-139.html#md138)

Maryland_Route_439

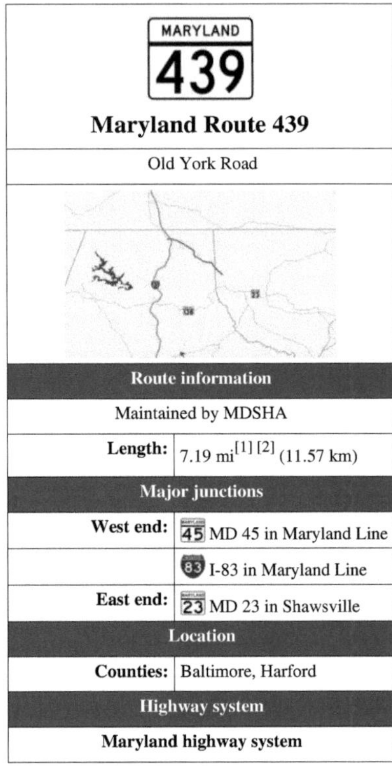

Maryland Route 439 (**MD 439**) is a state highway in the U.S. state of Maryland. Known as Old York Road, the state highway runs 7.19 miles (11.57 km) from MD 45 in Maryland Line east to MD 23 in Shawsville. In conjunction with MD 23 and Interstate 83 (I-83), MD 439 connects Bel Air with York, Pennsylvania. The state highway was constructed at both ends in the early 1930s. The middle section of MD 439 was brought into the state highway system around 1980.

Route description

MD 439 begins at an intersection with MD 45 (York Road) in Maryland Line a short distance south of the Pennsylvania state line. The state highway heads east through a four-ramp partial cloverleaf interchange with I-83 (Baltimore–Harrisburg Expressway). MD 439 heads southeast and passes through the hamlet of Shane as the highway follows the height of land between the drainage basins of Little Gunpowder Falls and Deer Creek.[1] [3] Shortly after crossing the Baltimore–Harford county line, the state highway reaches its eastern terminus at MD 23 (Norrisville Road) in Shawsville.[2] [3]

History

Old York Road was an alternate, less-direct route for traffic between Towson and Maryland Lane compared to York Road. MD 439 is one of several extant stretches of the road; another segment is from south of Shawsville to MD 145 in Jacksonville, part of which is MD 562.[3] MD 439 was paved as a state highway in two segments between 1930 and 1933: from York Road east to Lentz Road, and from the Baltimore–Harford county line east to MD 23.[4] [5] The state highway originally had an acute intersection with York Road north of the present terminal intersection. MD 439 was relocated when I-83 was completed from Parkton to the Pennsylvania state line in 1959.[6] The state highway remained in pieces until the intervening section of Old York Road was brought under state maintenance around 1981.[7]

Junction list

County	Location	Mile [1] [2]	Destinations	Notes
Baltimore	Maryland Line	0.00	45 MD 45 (York Road) – Parkton, Shrewsbury	Western terminus
		0.24	83 I-83 (Baltimore–Harrisburg Expressway) – Baltimore, York	I-83 Exit 36
Harford	Shawsville	7.19	23 MD 23 (Norrisville Road) – Norrisville, Jarrettsville	Eastern terminus
1.000 mi = 1.609 km; 1.000 km = 0.621 mi				

References

[1] "Highway Location Reference: Baltimore County" (http://www.roads.maryland.gov/Location/2010_BALTIMORE.pdf) (PDF). Maryland State Highway Administration. 2010. . Retrieved 2011-07-22.

[2] "Highway Location Reference: Harford County" (http://www.roads.maryland.gov/Location/2010_HARFORD.pdf) (PDF). Maryland State Highway Administration. 2010. . Retrieved 2011-07-22.

[3] Google, Inc. *Google Maps – Maryland Route 439* (http://maps.google.com/maps?saddr=MD-439+E/Old+York+Rd&daddr=MD-439+E/Old+York+Rd&hl=en&sll=39.703984,-76.644423&sspn=0.007809,0.01929&geocode=FUTZXQIdVHRu-w;FaviXAIdAuFv-w&mra=ls&t=h&z=12) (Map). Cartography by Google, Inc. . Retrieved 2011-07-22.

[4] *Report of the State Roads Commission of Maryland* (http://www.archive.org/details/reportofstateroa1927mary) (1927–1930 ed.). Baltimore: Maryland State Roads Commission. 1930-10-01. p. 199. . Retrieved 2011-07-22.

[5] Maryland Geological Survey. *Map of Maryland Showing State Road System: State Aid Roads and Improved County Road Connections* (Map) (1933 ed.).

[6] Maryland State Roads Commission. *Maryland: Official Highway Map* (Map) (1959 ed.).

[7] Maryland State Highway Administration. *Maryland: Official Highway Map* (Map) (1981-82 ed.).

External links

- MDRoads: MD 439 (http://www.mdroads.com/routes/420-439.html#md439)

Maryland_Route_136

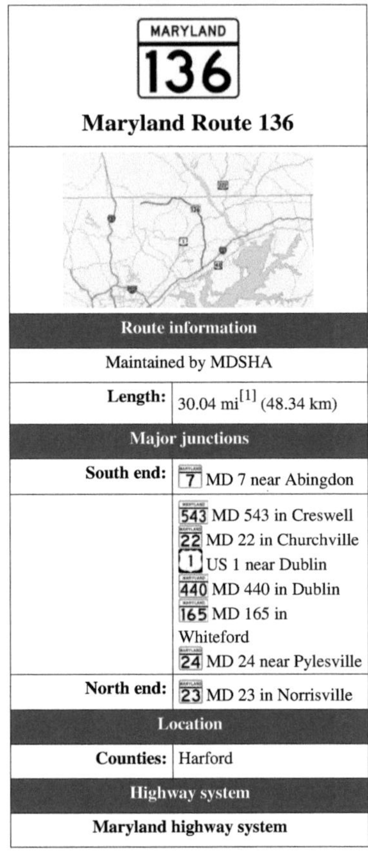

Maryland Route 136 (**MD 136**) is a state highway in the U.S. state of Maryland. The state highway runs 30.04 miles (48.34 km) from MD 7 near Abingdon north to MD 23 in Norrisville. MD 136 is an L-shaped route that connects the communities of Creswell, Churchville, Dublin, and Whiteford in eastern Harford County with each other and with Norrisville in the county's northwestern corner. The state highway is connected to the cities of Aberdeen and Havre de Grace via its connection with MD 22. MD 136 is also linked to the county seat of Bel Air from the east through MD 22, from the northeast by U.S. Route 1 (US 1), from the north via MD 24, and from the northwest by MD 23. The state highway starts on the coastal plain near the Chesapeake Bay and crosses Harford County's two main tributaries of the Susquehanna River, Deer Creek and Broad Creek, while traversing a wide swath of the Piedmont. MD 136 is the second longest Maryland state highway entirely within one county after MD 235.

The first section of MD 136 was constructed east from what is now MD 165 in Whiteford in the early 1920s. That highway was extended south to US 1 near Dublin in the mid 1920s. MD 136 was extended south to US 40 (now MD 7) in the early 1930s; the highway was also extended north from Whiteford to Graceton in the same period. The portion of MD 136 between MD 24 and MD 23 was originally MD 517, which was constructed in the mid 1930s. MD 136 was extended west over the gap between Graceton and MD 24 and assumed all of MD 517 to Norrisville in the mid 1950s.

Route description

MD 136 begins at an intersection with MD 7 (Philadelphia Road) east of Abingdon at the edge of Bush Declaration Natural Resources Management Area, just north of the confluence of Bynum Run and James Run to form the Bush River. The junction is near the site of the signing of the Bush Declaration, a 1775 resolution of Harford County citizens that supported the activities leading up to American Revolution. MD 136 heads north through farmland as two-lane Calvary Road, which crosses over Interstate 95 (John F. Kennedy Memorial Highway) with no access. I-95 can be accessed via MD 543 (Fountain Green Road/Creswell Road), which MD 136 intersects in the village of Creswell.[1] [2]

North of Creswell, MD 136 passes to the west of Stoney Demonstration Forest, crosses Broad Run adjacent to a quarry in the hamlet of Calvary, and passes close to the historic home Webster's Forest, which is accessed via Asbury Road. The state highway continues north to Churchville, where MD 136 intersects MD 22 (Churchville Road) next to the village's namesake, Churchville Presbyterian Church. A short distance east of the intersection, MD 22—which connects Bel Air and Aberdeen—intersects MD 155 (Level Road), which leads to Havre de Grace.[1] [2]

MD 136 leaves Churchville as Priestford Road, which passes by the Churchville Test Area, an auxiliary unit of Aberdeen Proving Ground used to test Army vehicles. The state highway traverses Deer Creek at Priest Ford, which is the site of Priest Neal's Mass House and Mill Site, a historic church, and a junction with Harmony Church Road, which leads through the Lower Deer Creek Valley Historic District. MD 136 passes a loop of old alignment, Poplar Grove Road, before intersecting US 1 (Conowingo Road) at an acute angle in the hamlet of Poplar Grove. The state highway continues north as Whiteford Road, which passes another loop of old alignment, Dublin School Road, before meeting MD 440 (Dublin Road) in the village of Dublin.[1] [2]

MD 136 heads northwest out of Dublin, crossing Broad Creek and intersecting Robinson Mill Road, which follows the creek east to the Broad Creek Soapstone Quarries. The state highway intersects MD 646 (Prospect Road) in the hamlet of Prospect before approaching Whiteford. Whiteford and the neighboring village of Cardiff are part of the Whiteford-Cardiff Historic District, which preserve buildings from the 19th century when the area was a center of the slate industry. MD 136 expands to a four-lane undivided highway and intersects Old Pylesville Road, which serves as the main street of the villages, before intersecting MD 165 (Pylesville Road).[1] [2]

MD 136 reduces to two lanes as it heads west from Whiteford toward the hamlet of Graceton, where the state highway intersects MD 624 (Graceton Road) within 2000 feet (610 m) of the Pennsylvania state line. The state highway veers southwest and crosses Broad Run again before reaching Five Forks, where the state highway intersects Clermont Mill Road and MD 24 (Rocks Road). MD 136 heads west as Harkins Road, intersecting Fawn Grove Road in the hamlet of Harkins and traversing three tributaries of Deer Creek—Falling Branch, Big Branch, and Island Branch—before reaching its northern terminus at MD 23 (Norrisville Road) in Norrisville.[1] [2]

History

The first section of MD 136 to be constructed was Whiteford Road from Prospect to Whiteford by 1923.[3] The remainder of Whiteford Road south to US 1 was constructed between 1924 and 1927.[4] [5] Priestford Road was built starting in 1930 from US 1.[6] Both the Priestford Road and Calvary Road portions of MD 136 were completed in 1932, including a new bridge over Deer Creek at Priest Ford. In 1933, MD 136 achieved its original extent when the state highway was extended north from Whiteford to Graceton.[7] [8] The Harkins Road portion of MD 136 was originally designated MD 517.[9] MD 517 was constructed as a modern highway from MD 24 at Five Forks west to Harkins in 1933.[7] [8] Harkins Road was extended west 2 miles (3.2 km) from Harkins in 1934 and 1935.[7] [10] MD 517 was completed west to MD 23 in Norrisville in 1936.[11] [12] MD 136 was relocated through Dublin in 1952 and 1953.[13] The state highway reached its current extent in 1956 when the county highway between Graceton and Five

Forks was transferred to state control; MD 136 was extended southwest to Five Forks then assumed all of MD 517 to its present northern terminus in Norrisville.[14]

Junction list

The entire route is in Harford County.

Location	Mile[1]	Destinations	Notes
Abingdon	0.00	**7** MD 7 (Philadelphia Road) – Riverside, Edgewood	Southern terminus
Creswell	2.39	**543** MD 543 (Fountain Green Road/Creswell Road) – Riverside, Fountain Green	
Churchville	6.19	**22** MD 22 (Churchville Road) – Bel Air, Aberdeen, Havre de Grace	
Dublin	11.81	**1** US 1 (Conowingo Road) – Bel Air, Rising Sun	
	12.81	**440** MD 440 (Dublin Road) – Ady, Darlington	
Prospect	16.24	**646** MD 646 south (Prospect Road) / Prospect Road north – Ady	
Whiteford	19.16	**165** MD 165 (Pylesville Road) – Pylesville, Cardiff	
Graceton	21.64	**624** MD 624 (Graceton Road) – Pylesville, Fawn Grove, PA	
	23.75	**24** MD 24 (Rocks Road) – Bel Air, Fawn Grove, PA	
Norrisville	30.04	**23** MD 23 (Norrisville Road) – Jarrettsville, Stewartstown, PA	Northern terminus
1.000 mi = 1.609 km; 1.000 km = 0.621 mi			

References

[1] "Highway Location Reference: Harford County" (http://www.marylandroads.com/Location/2009_HARFORD.pdf) (PDF). Maryland State Highway Administration. 2009. . Retrieved 2011-02-23.

[2] Google, Inc. *Google Maps – Maryland Route 136* (http://maps.google.com/maps?f=d&source=s_d&saddr=MD-136+N/Calvary+Rd& daddr=MD-136+N/Whiteford+Rd+to:MD-136+N/Harkins+Rd&hl=en& geocode=FZxSWgIdAk10-w;FZYOXQIdUjd0-w;FSbFXQId9jRw-w&mra=ls&sll=39.652689,-76.26503&sspn=0.007467,0.01929& ie=UTF8&t=h&z=10) (Map). Cartography by Google, Inc. . Retrieved 2011-02-23.

[3] Maryland Geological Survey. *Map of Maryland: Showing State Road System and State Aid Roads* (Map) (1923 ed.).

[4] *Report of the State Roads Commission of Maryland* (http://www.archive.org/details/annualreportsofs1924mary). **1924-1926**. Baltimore: Maryland State Roads Commission. January 1927. pp. 44–45, 84. . Retrieved 2011-02-23.

[5] Maryland Geological Survey. *Map of Maryland: Showing State Road System and State Aid Roads* (Map) (1927 ed.).

[6] *Report of the State Roads Commission of Maryland* (http://www.archive.org/details/reportofstateroa1927mary). **1927-1930**. Baltimore: Maryland State Roads Commission. 1930-10-01. p. 214. . Retrieved 2011-02-23.

[7] *Report of the State Roads Commission of Maryland* (http://www.archive.org/details/reportofstateroa1931mary). **1931-1934**. Baltimore: Maryland State Roads Commission. 1934-12-28. pp. 338–339. . Retrieved 2011-02-23.

[8] Maryland Geological Survey. *Map of Maryland Showing State Road System: State Aid Roads and Improved County Road Connections* (Map) (1933 ed.).

[9] Maryland State Roads Commission. *General Highway Map: State of Maryland* (Map) (1939 ed.).

[10] Maryland Geological Survey. *Map of Maryland Showing State Road System: State Aid Roads and Improved County Road Connections* (Map) (1935 ed.).

[11] *Report of the State Roads Commission of Maryland* (http://www.archive.org/details/reportofstateroa1935mary). **1935-1936**. Baltimore: Maryland State Roads Commission. 1936-12-04. p. 81. . Retrieved 2011-02-23.

[12] Maryland State Roads Commission. *Map of Maryland Showing State Road System* (Map) (1936 ed.).

[13] *Report of the State Roads Commission of Maryland* (http://www.archive.org/details/reportofstateroa1953mary). **1953-1954**. Baltimore: Maryland State Roads Commission. 1954-11-12. p. 179. . Retrieved 2011-02-23.

[14] Maryland State Roads Commission. *Maryland: Official Highway Map* (Map) (1956 ed.).

External links

- MD 136 @ MDRoads.com (http://www.mdroads.com/routes/120-139.html#md136)

Ma_and_Pa_Trail

Ma and Pa Trail	
Length	6.25 miles (10.1 km)
Location	Harford County, Maryland, USA
Trailheads	Williams Street, Equestrian Center Friends Park, Melrose Lane
Use	Walking/Biking/Riding
Trail difficulty	Easy
Season	All year
Sights	Heavenly Waters Park, Friends Park, Bel Air
Hazards	Grade Shifts, Tunnels

The **Ma and Pa Trail** is a 6.25-mile (10.1 km) multi-purpose rail trail that follows the path of the old Ma and Pa Railroad through Harford County, Maryland. It contains three linked trails, one unconnected, with plans to bridge the gap in the middle, bringing the total length to 10 miles (16 km). Both links of the trail are paved, have few steep hills or hazards, and contain overlooks and bridges when necessary.

The **Bel Air Trail** is a 2.5-mile (4.0 km) trail that runs from Williams Street in downtown to the Equestrian Center on the other side of the bypass. It passes underneath of the Route 24 Highway, the U.S. Route 1 bypass, and crosses Tollgate Road, where it meets the Fallston Trail. Spurs off of it connect to the Harford Mall and Liriodendron mansion, and branches out to a dog park at the end. It serves as a shortcut across town, rather than walking along highways, and contains several bridges across branches of the creek while cutting through the forested Heavenly Waters Park. Bathrooms and parking are available at each end of the trail. Horses and dogs are permitted on the Bel Air link of the path.

The **Fallston Link** is a 2-mile (3.2 km) segment dedicated on June 7, 2008, that runs from the Equestrian Center to the Edgeley Grove Farm area of Fallston.[1]

The **Forest Hill Trail** is a third, 1.75-mile (2.8 km) link of the trail running from the duckpond at Friends Park to Melrose Lane. A tunnel is located at Maryland Route 23, but passes over a development street on the north end. Bathrooms and parking are located at each end. Dogs are allowed on the Forest Hill link, but horses are not.

As of late 2006, planning is underway to link the two routes into one continuous trail . At the time of the original constructions, a tunnel was built underneath of Route 1 in preparation, but it is unconnected to any path system. Construction for an extension from Bel Air to the Edgeley Grove area in Fallston is planned, and a 4.1-mile (6.6 km) northern link at Dooley Road, near the state line in Cardiff is under study.

References

[1] "Bel Air trail chugs along". *The Baltimore Sun*. 2008-06-02. p. 2B.

External links

- Official Site (http://www.mapatrail.org/)

Deer_Creek_(Maryland)

Deer Creek is a 52.9-mile-long (85.1 km)[1] river in Maryland that flows through the scenic areas of Harford County and empties into the Susquehanna River, roughly halfway between the Interstate 95 bridge and Conowingo Dam. It serves as a divider between the agricultural and urban/suburban areas of Harford County.

Deer Creek rises in Shrewsbury in York County, Pennsylvania, and flows southeast, soon entering Maryland. It cuts across the northeastern corner of Baltimore County and into Harford County, where it runs through Rocks State Park and Susquehanna State Park, passing north of the Bel Air area.

Deer Creek supported the last known population of the Maryland darter, the only endemic vertebrate in Maryland [2]. The creek is commonly used for recreation in the summer months.

References

[1] U.S. Geological Survey. National Hydrography Dataset high-resolution flowline data. The National Map (http://viewer.nationalmap.gov/ viewer/), accessed August 8, 2011

[2] http://www.dnr.state.md.us/wildlife/mddarter.asp

Forest_Hill,_Maryland

Forest Hill, Maryland	
— Unincorporated community —	
Coordinates: 39°35′06″N 76°23′16″W	
Country	United States
State	Maryland
County	Harford
Elevation[1]	577 ft (176 m)
Time zone	Eastern (EST) (UTC-5)
– Summer (DST)	EDT (UTC-4)
ZIP code	21050
Area code(s)	410, 443
FIPS code	24-24025
GNIS feature ID	584446[1]
Other name	Foresthill[1]
[2]	

Forest Hill is an unincorporated community in Harford County, Maryland, United States located north of the county seat of Bel Air. The main part of town is located at the intersection of Maryland Route 24 and Jarrettsville Road (former Maryland Route 23). Until 1958, this community was served by the Maryland and Pennsylvania Railroad at milepost 30.3.

Forest Hill's zip code area covers a relatively large area, with rural land on one side and suburban neighborhoods on the other. The latter is part of the Bel Air suburbs.

St. Ignatius Church was listed on the National Register of Historic Places in 1974.[3]

Geography

Forest Hill is located at 39°35′06″N 76°23′16″W (39.585106, -76.387739).[1] Its elevation is 577 feet (176 m).

Demographics

At the 2000 census there were 14,951 people(7,234 men and 7,717 women) and 5,459 housing units in the Forest Hill area. The town's land area is 29.89 square miles (77.4 km^2).

References

[1] U.S. Geological Survey Geographic Names Information System: Forest Hill, Maryland (http://geonames.usgs.gov/pls/gnispublic/ f?p=gnispq:3:::NO::P3_FID:584446). Retrieved on 2008-07-10.

[2] "Forest Hill MD" (http://www.zipinfo.com/cgi-local/zipsrch.exe?cnty=cnty&ac=ac&zip=21050 &Go=Go). ZIP Code Lookup. . Retrieved 2008-07-10.

[3] "National Register Information System" (http://nrhp.focus.nps.gov/natreg/docs/All_Data.html). *National Register of Historic Places*. National Park Service. 2008-04-15. .

Jarrettsville,_Maryland

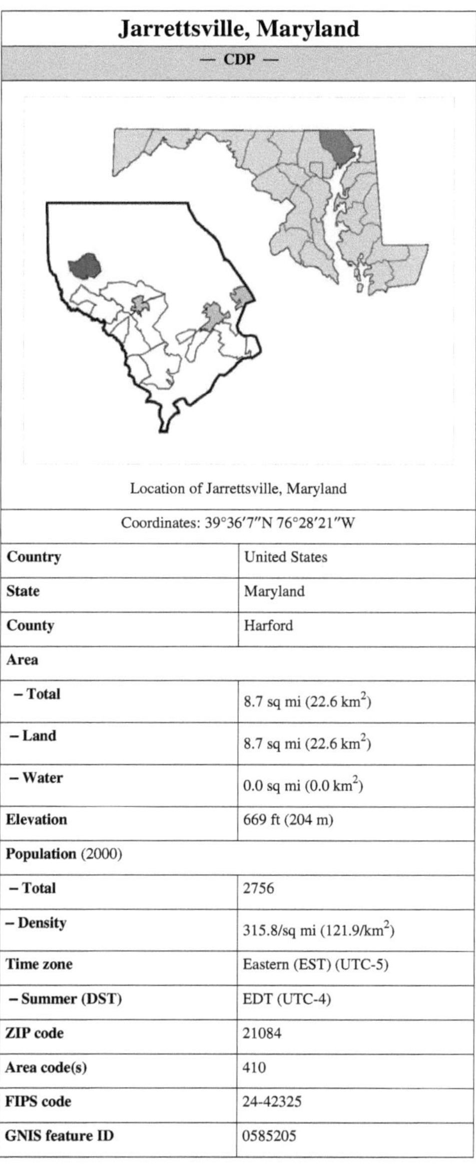

Jarrettsville, Maryland	
— CDP —	
Location of Jarrettsville, Maryland	
Coordinates: 39°36′7″N 76°28′21″W	
Country	United States
State	Maryland
County	Harford
Area	
– Total	8.7 sq mi (22.6 km^2)
– Land	8.7 sq mi (22.6 km^2)
– Water	0.0 sq mi (0.0 km^2)
Elevation	669 ft (204 m)
Population (2000)	
– Total	2756
– Density	315.8/sq mi (121.9/km^2)
Time zone	Eastern (EST) (UTC-5)
– Summer (DST)	EDT (UTC-4)
ZIP code	21084
Area code(s)	410
FIPS code	24-42325
GNIS feature ID	0585205

Jarrettsville is a census-designated place (CDP) in Harford County, Maryland, United States. The population was 2,756 at the 2000 census.

History

My Lady's Manor was listed on the National Register of Historic Places in 1978, and includes portions of Jarrettsville.[1] The town was named for the Jarrett family, who farmed the area during the 1800s.

Geography

Jarrettsville is located at 39°36'7"N 76°28'21"W (39.601954, -76.472404)[2] .

According to the United States Census Bureau, the CDP has a total area of 8.7 square miles (23 km^2), all of it land.

Demographics

As of the census[3] of 2000, there were 2,756 people, 900 households, and 781 families residing in the CDP. The population density was 315.8 people per square mile (121.9/km²). There were 918 housing units at an average density of 105.2/sq mi (40.6/km²). The racial makeup of the CDP was 97.21% White, 1.16% African American, 0.15% Native American, 0.29% Asian, 0.33% from other races, and 0.87% from two or more races. Hispanic or Latino of any race were 0.54% of the population.

There were 900 households out of which 43.7% had children under the age of 18 living with them, 77.6% were married couples living together, 6.4% had a female householder with no husband present, and 13.2% were non-families. 11.1% of all households were made up of individuals and 5.4% had someone living alone who was 65 years of age or older. The average household size was 3.05 and the average family size was 3.30.

In the CDP the population was spread out with 28.9% under the age of 18, 6.5% from 18 to 24, 26.4% from 25 to 44, 29.0% from 45 to 64, and 9.2% who were 65 years of age or older. The median age was 39 years. For every 100 females there were 101.3 males. For every 100 females age 18 and over, there were 97.3 males.

The median income for a household in the CDP was $69,632, and the median income for a family was $81,771. Males had a median income of $51,524 versus $31,905 for females. The per capita income for the CDP was $29,246. None of the families and 1.4% of the population were living below the poverty line, including no under eighteens and 11.6% of those over 64.

Noted natives and residents

- Randy McMillan -- Former NFL running back with the Baltimore/Indianapolis Colts.
- Mel Kiper Jr. -- ESPN NFL draft day analyst.
- Cyrus Chestnut -- jazz pianist.

References

[1] "National Register Information System" (http://nrhp.focus.nps.gov/natreg/docs/All_Data.html). *National Register of Historic Places*. National Park Service. 2008-04-15. .
[2] "US Gazetteer files: 2010, 2000, and 1990" (http://www.census.gov/geo/www/gazetteer/gazette.html). United States Census Bureau. 2011-02-12. . Retrieved 2011-04-23.
[3] "American FactFinder" (http://factfinder.census.gov). United States Census Bureau. . Retrieved 2008-01-31.

Norrisville,_Maryland

Norrisville is an unincorporated community in Harford County, Maryland, just south of the Pennsylvania state line in the extreme northwest section of the county. It is primarily a farming community full of rolling hills and small valleys, and is part of the Deer Creek watershed. The area is served by two state roads, Maryland routes 23 and 136. A demographic profile estimates that the Norrisville area had a population of 2,931 as of 2000. It lies at an elevation of 741 feet (226 m). Norrisville is part of the White Hall zip code 21161, whose post office is located just over the Baltimore County line.

References

- U.S. Geological Survey Geographic Names Information System: Norrisville, Maryland [1]

References

[1] http://geonames.usgs.gov/pls/gnispublic/f?p=gnispq:3:::NO::P3_FID:586209

Maryland_and_Pennsylvania_Railroad

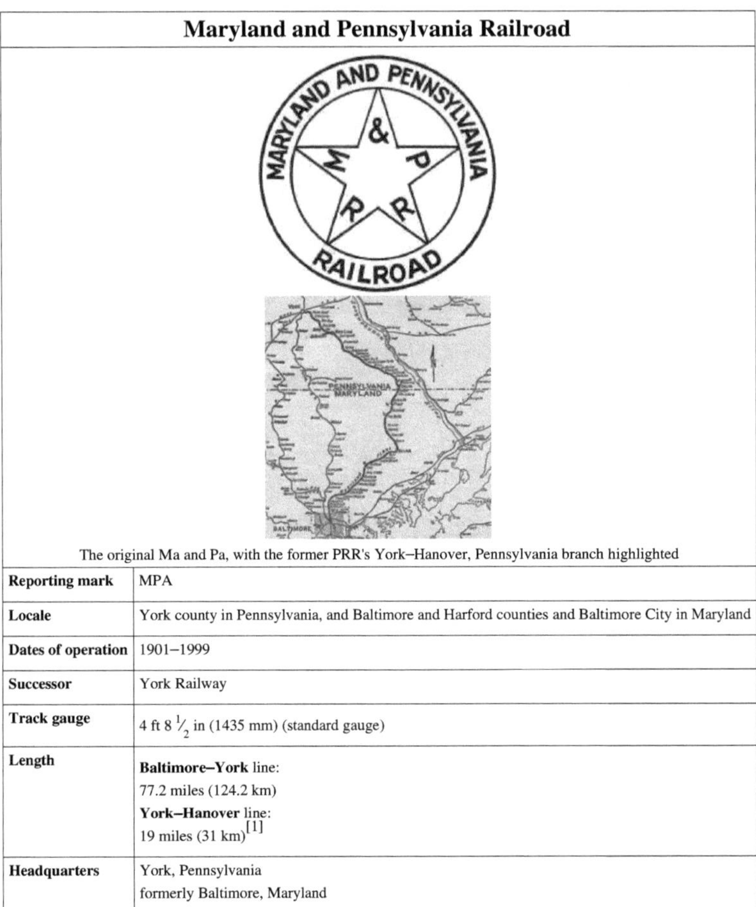

Maryland and Pennsylvania Railroad	
The original Ma and Pa, with the former PRR's York–Hanover, Pennsylvania branch highlighted	
Reporting mark	MPA
Locale	York county in Pennsylvania, and Baltimore and Harford counties and Baltimore City in Maryland
Dates of operation	1901–1999
Successor	York Railway
Track gauge	4 ft 8 $\frac{1}{2}$ in (1435 mm) (standard gauge)
Length	**Baltimore–York** line: 77.2 miles (124.2 km) **York–Hanover** line: 19 miles (31 km)[1]
Headquarters	York, Pennsylvania formerly Baltimore, Maryland

The **Maryland and Pennsylvania Railroad** (reporting mark **MPA**), familiarly known as the **"Ma and Pa"**, was an American short-line railroad between York and Hanover, Pennsylvania, formerly operating passenger and freight trains on its original line between York and Baltimore, Maryland, from 1901 until the 1950s. The Ma and Pa was popular with railfans in the 1930s and 1940s for its antique equipment and curving, picturesque right-of-way through the hills of rural Maryland and Pennsylvania. Reflecting its origin as the unintended product of the merger of two 19th-century narrow gauge railways, the meandering Ma and Pa line took 77.2 miles (124 km) to connect Baltimore and York, although the two cities are only 45 miles (72 km) apart in a straight line.[2]

Passenger service was discontinued on August 31, 1954, and the section from Baltimore to Whiteford, Maryland (just south of the Mason-Dixon line demarcating the Pennsylvania-Maryland border) was abandoned in June 1958. Most of the remaining original railroad line was abandoned by 1984. The Maryland and Pennsylvania Railroad acquired a former 19-mile (31 km) Pennsylvania Railroad (PRR) branch line between York and Hanover in the 1980s, now operated by a successor corporation, York Railway.[3]

History

19th-century predecessors

The Maryland and Pennsylvania Railroad was formed from two earlier 19th-century 3-foot (914 mm) narrow gauge railways: the Baltimore & Delta Railway, later the **Baltimore & Lehigh Railway**, and the York & Peach Bottom Railway, later the **York Southern Railroad**.[4] Construction of the Baltimore & Delta Railway started in 1881, and passenger trains between Baltimore and Towson, Maryland began on April 17, 1882, extended to Bel Air, Maryland on June 21, 1883.[5] The following January, the line was completed to Delta, Pennsylvania.

In Pennsylvania, the York & Peach Bottom Railway was incorporated in 1871, laying 3-foot (914 mm) gauge track between York and Red Lion by August 1874 and completing its line to Delta in 1876.[6] Both railroads struggled with light freight traffic and financial difficulties in the 1890s. Because of their narrow gauge construction, the Baltimore & Lehigh Railway and York Southern Railroad could not interchange freight cars with other lines.[2] Finally, the Baltimore & Lehigh Railway and York Southern Railroad converted to standard gauge between 1898–1900 and subsequently merged to form the Maryland and Pennsylvania Railroad on February 12, 1901.[4] The result was the circuitous, 77.2-mile (124 km) "Ma and Pa" route between Baltimore and York, compared to the competing Pennsylvania Railroad's more direct 56-mile (90 km) distance between the two cities on its Northern Central Railway division.[2] The completed line had grades up to 2.3 percent and 55 sharp curves of 16–20 degrees (most mainline railroads seldom exceed six degrees, and even the former Denver and Rio Grande Western Railroad's mainline through the Rocky Mountains does not exceed 12 degrees).[2]

20th century

Following the merger, the Ma and Pa operated through passenger and freight trains between York and Baltimore, as well as local trains at each end of the line, hauling mail and express, slate, marble, anthracite coal, lumber, furniture, and agricultural products to market.[2] Particularly on the Pennsylvania Division (Delta–York), slate from Delta and manufactured goods from Red Lion and York were mainstays of the railroad's outbound freight traffic in the early years.[7] On the Maryland Division, inbound anthracite coal deliveries accounted for a significant volume of carloadings, along with milk from the many dairy farms in the area.[7] [8] One early morning train from Fallston boarded more than 1,100 gallons of milk daily and was dubbed the "Milky Way".[8] The line was profitable and traffic volume was such that additional locomotives were necessary.

Former Ma and Pa Station in Red Lion, Pennsylvania

Baldwin 0-6-0 locomotive #30, built in 1913 and owned by the Ma & Pa until 1956, was typical of the line's aged equipment.

The Ma and Pa acquired two 0-6-0 Baldwin switchers in 1913, #29 and #30 *(pictured)*, called "jewels of engines, in some respects the most attractive the road had", by writer George Hilton in *The Ma & Pa – A History of the Maryland & Pennsylvania Railroad*.[9] The next year, three 2-8-0 "Consolidations" by Baldwin were added to the roster, providing more powerful locomotives for the Baltimore–York through freights.[9] At its peak, the railroad had 16 locomotives and 160 pieces of rolling stock, with 573 employees.[10]

With increasing competition from trucks and automobiles in the 1920s, passenger volume began to decline along with less-than-carload freight, such as milk from the many dairy farms along the Ma and Pa's pastoral route. The Ma and Pa substituted more economical, self-propelled gas-electric passenger cars for steam-powered passenger trains in 1927–1928. Carload freight volume increased in the 1920s, however, as more industries located along the line, and earnings were strong enough for the company to declare dividends in 1930 and 1931.[11] The Ma and Pa's relative prosperity ended with the economic downturn during the Great Depression, which cut the railroad's gross revenues by half from 1932 to 1935.[12]

In the mid-1930s, the Ma and Pa became an early favorite of railfans, attracted by its hilly, curving line through rural Maryland and Pennsylvania. The railroad offered several popular fan excursions pulled by its elderly steam locomotives.[13]

Following the end of World War II, the Ma and Pa acquired four diesel locomotives for more economical operations, but traffic declined significantly. When the Ma and Pa's mail contract was cancelled by the U.S. postal service, the railroad discontinued all passenger service on August 31, 1954.[14] One person on the last passenger train recalled that many riders came from as far away as Boston, Massachusetts, and Washington, D.C., to participate in the historic event, along with members of the National Railway Historical Society.[15] The picturesque line's last steam engine dropped its fire for the final time on November 29, 1956.[10]

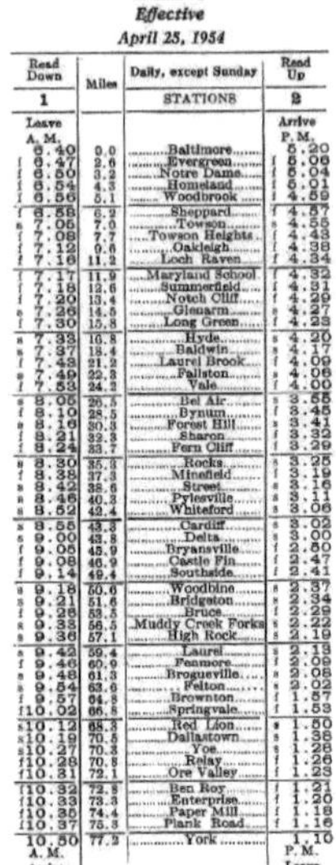

TIME TABLE

EASTERN STANDARD TIME

Effective
April 25, 1954

Read Down 1	Miles	STATIONS	Read Up 2
Leave A. M.			Arrive P. M.
6.40	0.0	Baltimore	5.20
f 6.47	2.6	Evergreen	f 5.06
f 6.50	3.2	Notre Dame	f 5.04
f 6.54	4.3	Homeland	f 5.01
f 6.56	5.1	Woodbrook	f 4.59
f 6.58	6.2	Sheppard	f 4.57
s 7.05	7.0	Towson	s 4.55
f 7.08	7.7	Towson Heights	f 4.43
f 7.12	9.6	Oakleigh	f 4.38
f 7.16	11.2	Loch Raven	f 4.34
f 7.17	11.9	Maryland School	f 4.32
f 7.18	12.6	Summerfield	f 4.31
f 7.20	13.4	Notch Cliff	f 4.29
s 7.26	14.5	Glenarm	s 4.27
f 7.30	15.8	Long Green	f 4.23
s 7.33	16.8	Hyde	s 4.20
s 7.37	18.4	Baldwin	s 4.17
f 7.43	21.2	Laurel Brook	f 4.09
s 7.49	22.3	Fallston	s 4.08
f 7.53	24.2	Vale	f 4.00
s 8.05	26.5	Bel Air	s 3.55
f 8.10	28.5	Bynum	f 3.45
s 8.16	30.3	Forest Hill	s 3.41
f 8.21	32.3	Sharon	f 3.32
f 8.24	33.7	Fern Cliff	f 3.29
s 8.30	35.3	Rocks	s 3.25
f 8.38	37.3	Minefield	f 3.19
s 8.42	38.6	Street	s 3.16
s 8.46	40.3	Pylesville	s 3.11
s 8.52	42.4	Whiteford	s 3.06
s 8.55	43.3	Cardiff	s 3.02
s 9.00	43.8	Delta	s 3.00
f 9.05	45.9	Bryansville	f 2.50
f 9.08	46.9	Castle Fin	f 2.47
f 9.14	49.4	Southside	f 2.41
s 9.18	50.6	Woodbine	s 2.37
s 9.21	51.6	Bridgeton	s 2.34
f 9.26	53.5	Bruce	f 2.29
s 9.33	56.5	Muddy Creek Forks	s 2.22
s 9.36	57.1	High Rock	s 2.19
s 9.42	59.4	Laurel	s 2.13
f 9.46	60.9	Fenmore	f 2.09
s 9.48	61.3	Brogueville	s 2.08
s 9.54	63.6	Felton	s 2.02
f 9.57	64.8	Brownton	f 1.57
f 10.02	66.8	Springvale	f 1.53
s 10.12	68.3	Red Lion	s 1.50
s 10.19	70.5	Dallastown	s 1.38
s 10.27	70.3	Yoe	s 1.28
f 10.28	70.8	Relay	f 1.26
f 10.31	72.1	Ore Valley	f 1.23
f 10.32	72.5	Ben Roy	f 1.21
f 10.33	73.3	Enterprise	f 1.20
f 10.35	74.4	Paper Mill	f 1.18
f 10.37	75.3	Plank Road	f 1.16
10.50 A. M. Arrive	77.2	York	1.10 P. M. Leave

f—Flag Station.   s—Stop.
Trains will stop at stations marked "f" on signal or notice to Conductor. Presentation of tickets to these stations will be considered sufficient notice.

Final "Ma and Pa" passenger
timetable, 1954

The last "Ma and Pa" train departs Towson, Maryland, on June 11, 1958

The lack of traffic on the railroad's 44-mile (71 km) Baltimore–Whiteford Maryland Division in the 1950s was particularly acute. One of the last major shipments to occur was Indiana limestone for the construction of Baltimore's Cathedral of Mary Our Queen in 1956.[16] The Baltimore–Whiteford segment in Maryland was finally abandoned altogether on June 11, 1958, leaving only the stone abutments where the tracks crossed York Road in Towson on a steel girder bridge. A local group of history buffs placed a bronze plaque on the west abutment in 1999, commemorating the departed railroad's place in Towson history.[17]

In the 1960s, the Ma and Pa Railroad continued to solicit business along its line for its remaining 34.8-mile (56.0 km) Whiteford–York segment, almost entirely in Pennsylvania. In 1964, it added a siding 905 feet (276 m) long near Red Lion to serve a new cigar box factory.[18] In 1971, the Maryland and Pennsylvania Railroad was acquired by Emons Industries.[19] Primarily hauling slate from a quarry at Delta, and furniture from a factory in Red Lion, the Ma and Pa's Pennsylvania Division continued in operation until June 14, 1978, when the line was further reduced to the 9-mile (14 km) York–Red Lion section.[10] The Red Lion freight station was closed on November 1, 1980; when the Pennsylvania town's furniture manufacturer shuttered its doors in 1984, the Red Lion section of the railroad was also abandoned.[3] [10]

Currently

The Emons-controlled Maryland and Pennsylvania Railroad acquired 19 miles (31 km) of a former Pennsylvania Railroad (PRR) branch line between York and Hanover, Pennsylvania in the 1980s.[3] In December 1999, Emons merged its Maryland and Pennsylvania Railroad subsidiary with another area short-line, Yorkrail, forming the York Railway.[20] In 2002, Genesee and Wyoming gained control of the 42-mile (68 km) York Railway, including the former Maryland and Pennsylvania Railroad trackage between York–Hanover.[19] The York Railway currently serves 40 online rail customers and connects with the Norfolk Southern and CSX railroads.[21]

Locomotive #81, an EMD NW2 acquired in 1946, now at the Railroad Museum of Pennsylvania

A small, 3-mile (5 km) fragment of the original railroad line still exists between Laurel and Muddy Creek Forks in York County, Pennsylvania, maintained by the Maryland and Pennsylvania Railroad Preservation Society.[3] [8] Founded by enthusiasts and former employees in 1986, the group has restored the Muddy Creek Forks station and also has a small collection of rolling stock there.[8] The preserved Red Lion station is now a museum operated by the Red Lion Area Historical Society.[10] Another section of the Ma and Pa's old right-of-way was converted in 1998 to a rail trail in Harford County, Maryland. Now 6 miles (10 km) long, the MA & PA Heritage Trail through Bel Air is used for hiking and biking.[22]

In Baltimore, near Pennsylvania Station, Ma and Pa track remnants and the old roundhouse, freight shed, and yard shed are still extant.[23] The Baltimore Streetcar Museum now operates in this area.

See also

- Maryland and Pennsylvania Railroad Preservation Society [24]
- Film of steam-powered excursion train on the Ma & Pa, powered by Canadian Pacific 4-6-2 engine #1286. [25]
- Ghosts of the Maryland & Pennsylvania Railroad (Baltimore area) [26]
- "Ma and Pa" Railroad in Towson, Md.

References

[1] *Official Guide of the Railways*. New York: National Railway Publication Co., February 1956, p. 347.

[2] Frank P. Donovan, editor (1949). *"The Ma & Pa by William Moedinger Jr."*. *Railroads of America*. Milwaukee, WI: Kalmbach Publishing. LCCN 49-048570.

[3] Craig Sansonetti (1997-06-05). "A History of the Maryland & Pennsylvania Railroad" (http://web.archive.org/web/20051230233324/ http://www.maandparailroad.com/mapahistory.html). Maryland and Pennsylvania Railroad Preservation Society. Archived from the original (http://www.maandparailroad.com/mapahistory.html) on December 30, 2005. . Retrieved 2008-05-25.

[4] George W. Hilton (1963). *The Ma & Pa — A History of the Maryland & Pennsylvania Railroad*. Berkeley, CA: Howell-North Books. p. 61. LCCN 63-017444.

[5] Hilton, p. 22.

[6] Hilton, p. 11.

[7] Hilton, p. 83.

[8] Allen, Bob (October 15, 2003). "Recalling the little engine that could". *The Towson Times*: p. 34.

[9] Hilton, p. 87.

[10] "The Red Lion Train Station History" (http://www.redlionpa.org/history.htm). Red Lion (Pa.) Area Historical Society. 2007-06-25. . Retrieved 2008-07-07.

[11] Hilton, p. 111.

[12] Hilton, p. 114.

[13] Herbert H. Harwood, Jr. (2004–2005). "Maryland & Pennsylvania Railroad: The 'Ma & Pa'" (http://www.mdoe.org/md_penn_railroad. html). Maryland Online Encyclopedia. . Retrieved 2008-07-07.

[14] Hilton, p. 141.

[15] John R. Eicker (August 30, 1964). "The Ma and Pa's Last Run from Baltimore to York". *The Baltimore Sun.*

[16] Hilton, p. 151.

[17] Loni Ingraham (May 26, 1999). "'Ma and Pa' railroad abutments get HTI plaque". *The Towson Times.*

[18] Paul Kulishek (Spring 2008). "White Flag Extra". *Timetable* (Maryland and Pennsylvania Railroad Historical Society) **23** (2): 16.

[19] "About the Maryland and Pennsylvania Railroad" (http://www.maparailroadhist.org/history.htm). The Maryland and Pennsylvania Railroad Historical Society. 2008-01-25. . Retrieved 2008-05-30.

[20] "Emons Transportation kicks off new growth plan" (http://www.highbeam.com/doc/1G1-59363032.html). *Railway Age.* 2000-01-01. . Retrieved 2008-05-30.

[21] "York Railway Company" (http://www.gwrr.com/default.cfm?action=rail§ion=3B16a). Genesee and Wyoming Inc. . Retrieved 2008-05-30.

[22] "Bel Air trail chugs along". *The Baltimore Sun.* 2008-06-02. p. 2B.

[23] "Ghosts of the Maryland and Pennsylvania" (http://www.btco.net/ghosts/railroads/mpa/mapa.html). . Retrieved 13 May 2011.

[24] http://www.maandparailroad.com/

[25] http://www.youtube.com/watch?v=dfXymTYrNh0

[26] http://www.btco.net/ghosts/railroads/mpa/mapa.html

External links

- York Railway (http://www.gwrr.com/operations/railroads/north_america/york_railway), official Genesee and Wyoming website
- The Maryland and Pennsylvania Railroad Historical Society (http://www.maparailroadhist.org/history.htm)
- MA and PA Railroad Preservation Society (http://www.maandparailroad.com/index.html)
- Ma and Pa Railroad History in Harford County (http://www.mapatrail.org/rr.htm) (MA & PA Heritage Trail website)

Article Sources and Contributors

Maryland_Route_23 *Source*: http://en.wikipedia.org/w/index.php?title=Maryland_Route_23 *Contributors*: Algorerhythms, Doctor Whom, Dough4872, Ebyabe, Geoboe84, Imzadi1979, Jeff02, LilHelpa, Onore Baka Sama, PaulTanenbaum, Sebwite, TwinsMetsFan, Viridiscalculus, WOSlinker, Wildthing61476, 2 anonymous edits

U.S._Route_1_in_Maryland *Source*: http://en.wikipedia.org/w/index.php?title=U.S._Route_1_in_Maryland *Contributors*: Algorerhythms, Chaswmsday, Cpzilliacus, Deanlaw, Doctor Whom, Dough4872, Dwightrholmes, EJVargas, EikwaR, FlugKerl, Fredddie, Gfoley4, Imzadi1979, InTheAM, Jeff02, JustAGal, Jwsnottingham, Levineps, LilHelpa, Lwalt, Morriswa, NE2, O, Onore Baka Sama, Q300r bc2, Ravensfan5252, Sebwite, Tckma, TheOneKEA, TwinsMetsFan, Viridiscalculus, Wiki Raja, 27 anonymous edits

Pennsylvania_Route_24 *Source*: http://en.wikipedia.org/w/index.php?title=Pennsylvania_Route_24 *Contributors*: 8th Ohio Volunteers, Albertotineo10, Choess, Dough4872, Geoboe84, Golbez, Imzadi1979, Jjc104, Jllm06, O, Son, Stemonitis, TwinsMetsFan, 4 anonymous edits

Maryland_Route_165 *Source*: http://en.wikipedia.org/w/index.php?title=Maryland_Route_165 *Contributors*: Algorerhythms, Dough4872, Geoboe84, Imzadi1979, Jeff02, Onore Baka Sama, PBagian, Thisisbossi, Viridiscalculus, WOSlinker, Wildthing61476, 1 anonymous edits

Maryland_Route_146 *Source*: http://en.wikipedia.org/w/index.php?title=Maryland_Route_146 *Contributors*: Algorerhythms, Doctor Whom, Dough4872, Godietzy, Imzadi1979, JGHowes, Khatru2, Onore Baka Sama, OrangePaw17, Viridiscalculus, WOSlinker, Wildthing61476, 15 anonymous edits

Maryland_Route_138 *Source*: http://en.wikipedia.org/w/index.php?title=Maryland_Route_138 *Contributors*: Algorerhythms, Dough4872, Imzadi1979, Onore Baka Sama, Tim1357, Viridiscalculus, WOSlinker

Maryland_Route_439 *Source*: http://en.wikipedia.org/w/index.php?title=Maryland_Route_439 *Contributors*: Algorerhythms, Dough4872, Geoboe84, Imzadi1979, Jeff02, Onore Baka Sama, Psbjames70, SatuSuro, Vegaswikian, Viridiscalculus, WOSlinker

Maryland_Route_136 *Source*: http://en.wikipedia.org/w/index.php?title=Maryland_Route_136 *Contributors*: Algorerhythms, Clev7, Dough4872, Geoboe84, Imzadi1979, Onore Baka Sama, Viridiscalculus, WOSlinker

Ma_and_Pa_Trail *Source*: http://en.wikipedia.org/w/index.php?title=Ma_and_Pa_Trail *Contributors*: Backspace, Bleakcomb, Circeus, Daveplot, Deanlaw, JGHowes, Jllm06, Patleahy, PaulTanenbaum, Scba, 9 anonymous edits

Deer_Creek_(Maryland) *Source*: http://en.wikipedia.org/w/index.php?title=Deer_Creek_%28Maryland%29 *Contributors*: Duttler, J Clear, Jllm06, Ken Gallager, Kmusser, Ruhrfisch, Scba, Sebwite, 1 anonymous edits

Forest_Hill,_Maryland *Source*: http://en.wikipedia.org/w/index.php?title=Forest_Hill%2C_Maryland *Contributors*: Beachlover7245, BeastmasterGeneral, Clev44, Daveplot, DennisIsMe, Droll, Ebyabe, Gene93k, Jllm06, Nyttend, Pubdog, 11 anonymous edits

Jarrettsville,_Maryland *Source*: http://en.wikipedia.org/w/index.php?title=Jarrettsville%2C_Maryland *Contributors*: Discospinster, Jllm06, John Cardinal, LibLord, Lwalt, Pearle, Pubdog, Ram-Man, Seth Ilys, Siebau01, VerruckteDan, 19 anonymous edits

Norrisville,_Maryland *Source*: http://en.wikipedia.org/w/index.php?title=Norrisville%2C_Maryland *Contributors*: Brandon7533, Carlossuarez46, Closedmouth, Jllm06, Ken Gallager, 2 anonymous edits

Maryland_and_Pennsylvania_Railroad *Source*: http://en.wikipedia.org/w/index.php?title=Maryland_and_Pennsylvania_Railroad *Contributors*: Antandrus, Canadian2006, Choess, Deanlaw, Ilikeartrock, JGHowes, LarryMorseDCOhio, Lost on belmont, Lucky 6.9, Mboverload, MisfitToys, Must...edit...wikipedia, Mvincec, NE2, PaulTanenbaum, Piledhigheranddeeper, Rjwilmsi, Rmarquet, Scanlan, Shortride, Slambo, Smallbones, Tony1, 3 anonymous edits

Image Sources, Licenses and Contributors

Printed by Books on Demand GmbH, Norderstedt / Germany